Organic Farming and Modern Technology in Agriculture

NEW INDIA PUBLISHING AGENCY
New Delhi-110 034

About the Authors

Dr. Amit Kumar Jain has more than fourteen years of professional experience of teaching, research and administration. Dr Jain joined IGNOU as Assistant Regional Director (ARD) IGNOU Regional Centre, Ahmedabad, Gujarat in 2011, and presently working at IGNOU RC Karnal, Haryana. Prior to this, he has served as Assistant Professor in Microbiology, J.V. College, Baraut, Uttar Pradcsh. He obtained Ph.D in 2005 and Doctor of Science (D.Sc) in 2015 from CCS University, Meerut. He has qualified ICAR-NET in Plant Physiology and Biochemistry. He has also acquired the degree of B.Ed, MA(Edu.), MA in Distance Education (MADE), Six PG Diplomas i.e PGDNBT, PGDDE, PGDESD, PGDRD, PGDIPR and PGDEMA through ODL mode. He is the Fellow of International and National level academic societies i.e. Linnean Society of London (U.K.), Society of Innovative Educationalist & Scientific Research Professional (FSIESRP) and Botanical Society India (FBS) as well as Life member of the Indian Science Congress Association (ISCA), Kolkata & Centre of Education Growth and Research (CEGR), New Delhi. He has earned various national and international awards. In 2018, he awarded Innovative Researcher and Dedicated Academician Award 2018 (International Award) by INSRPM Malaysia, Dr. Jain has published 5 Books, 48 Research Articles in national and international journals, and published 8 Book chapters edited by various authors. Dr. Jain has participated in a large number of national and international seminars, conferences, workshops, training and added programmes. He is the member of various committees of academic bodies of other organizations.

Prof. J.D.S. Panwar, got his Ph.D. in Plant Physiology from Indian Agricultural Research Institute (IARI), New Delhi.Joined IARI, Plant Physiology in 1971 and served the Institute in various capacities as Scientist. Senior Scientist, Principal Scientist and as Head, Division of CPBM at IIPR, Kanpur. Retired as Head, Division of Plant Physiology, IARI, New Delhi. He was engaged in research for 37 years, teaching 19 years and guided to M.Sc. and Ph.D. students.He worked on several projects as PI and published 05 books,more than hundred Research Articles in journals of national and international repute, and participated in 14 International conferences. He has international exposure to work with Professor J.E. Harper, at University of Illinois, Urbana Champagin, USA as UNDP/FAO fellow. Prof. Panwaris a fellow of ISPP and ISPRD 1993 onward, awarded as ISPP, G.S. Sirohi best paper publication award (2003). He served as expert/members of academic and administrative bodies at national level including various reputed Universities He leaded as Vice Chancellor, Mahatama Gandhi University, Meghalaya as well as Venketashwara Open University, Arunachal Pradesh. He has actively participated in the Vice Chancellor`s conferences organized by Association of Indian Universities, New Delhi in collaboration of different universities by Amity University, Delhi, Swami Vivekananda Birth Anniversary, New Delhi, GLA University, Mathura and Assam University, Dibrugarh. In 2016, he has been awarded as Excellence Education Award by Confederation of Indian University at Kolkata. He has also participated in round table discussion on GST organized by "Education Promotion Society for India (EPSI)" in 2017.

Organic Farming and Modern Technology in Agriculture

Amit Kumar Jain
J.D.S. Panwar

NEW INDIA PUBLISHING AGENCY
New Delhi-110 034

NEW INDIA PUBLISHING AGENCY
101, Vikas Surya Plaza, CU Block, LSC Market
Pitam Pura, New Delhi – 110 034, India
Email: info@nipabooks.com
Web: www.nipabooks.com
For customer assistance, please contact
Phone: + 91-11-27 34 17 17 Fax: + 91-11- 27 34 16 16
E-Mail: feedbacks@nipabooks.com

ISBN : 978-93-90591-37-4

Composed and Designed by NIPA

Acknowledgement

First and foremost praise thanks to the God, the Almighty for never ending grace, and blessings to pursue our thoughts.

Many thanks to our teachers, colleagues, friends and well-wishers spanning all over the sphere, inspiring us for writing this book. We extend our thanks to the authors of various books who helped us in updating of the text to the present status.

This entire tenure made us to compromise with many of our family commitments and whole heartedly thanks to our beloved parents and family members for imparting unstinted support and encouragement. Sacrifice of our family kids especially Aarna Jain and Prakhar Panwar can't be expressed in words, they have compromised with their calendars during writing this book.

We express our heartfelt thanks to our publisher for liberal use of the recent available expertise in the field of printing and publication of the book which enabled us for timely publication.

Our readers are welcomed for the suggestions for further improvement in the next edition.

"Dedicated to our much-loved parents and family"

Authors

Preface

Modern agriculture and food sysyems, including organic agriculture are undergoing a technological and structural modernization and are faced with a growing globalization. Organic Agriculture is a modern way of farming management, using limited amount of chemical treatments which have negative effects on the environment, human health or animal health. It contributes to environmental protection and helps biodiversity to increase, organic farming does not mean going back to traditional methods of farming. Organic farming takes the best of these and combines them with modern scientific knowledge. In this way, the farmers create a healthy balance between nature and farming, where crops and animals can grow and thrive.

A perspective on organic agriculture has been presented in the book based on the available information in a cogent and easily understandable manner. In the book, there are 8 chapters oriented towards organic agriculture, environmental, income aspects and modernization in farming practices. Initially, the book provides an overall status of organic farming followed by biofertilizers i.e. granular, liquid and vermicomposting. Different views on various aspects of integrated agricultural management systems, double income of farmers, agricultural technological innovations are furthermore discussed with a focus on technological innovations. Finally, a vision was discussed on transformational role of digital technologies in agriculture. Selected or cited literature is also presented for further reading. It is hoped that the book will serve as a good reference source for those interested in organic agriculture.

Researcher, teachers, academics and students in the agricultural field in particular will find this book to be of immense use.

The authors would welcome suggestions from the readers to improve in subsequent editions.

Authors

Contents

1

Organic Farming History, Present Status and Future

Sustainable development has caught the imagination and action across the world. Sustainable agriculture is necessary to attain the goal of sustainable development. The word "sustain," is derived from the Latin sustinere (sus-, from below and tenere, to hold), to keep in existence or maintain, implies long-term support or permanence. As it pertains to agriculture, sustainable describes "Farming systems that are capable of maintaining productivity and usefulness to society indefinitely. Such systems must be resource-conserving, socially supportive, commercially competitive, and environmentally sound". "**Sustainable agriculture is the successful management of resources for agriculture to satisfy changing human needs while maintaining or enhancing the quality of environment and conserving natural resources**" (FAO). All definitions of sustainable agriculture lay great emphasis on maintaining an agriculture growth rate, which can meet the demand for food of all living things without draining the basic resources. Over time, the International Alliance for sustainable agriculture and an increasing number of researchers, farmers, policy-makers and organizations worldwide have developed a definition that unifies many diverse elements into a widely adopted, comprehensive, working definition: **A sustainable agriculture is ecologically sound, economically viable, socially just and humane**. Sustainable agriculture has not been carried out in a prescriptive manner, with a variety of ideas and farming models aimed at the objective of growing more food (for profit) while also providing environmental and social benefits (Pretty and Bharucha, 2014; Garibaldi *et al.,* 2017; Plumecocq *et al.,* 2018). Garibaldi *et al.,* (2017) present a partial list of concepts that have been proposed as a way of achieving agricultural sustainability, and we add to this non-exhaustive list by including others from the wider literature (Table 1.1, also Gold, 2007 for a longer list of related terms). Many of these ideas, such as integrated pest management, agroforestry, and organic agriculture are now quite familiar, whilst others, such as precision farming and sustainable intensification, are becoming more common. All of these terms have influenced the policy landscape at a variety of scales, as policy-makers constantly look for the best way of encouraging the adoption of sustainable agriculture in practice.

Table 1.1: Concepts related to sustainable agriculture (partially from Garibaldi *et al.,* 2017).

Concept	Suggested definition (may vary between sources)	References
Agroecology	The study of ecological processes, particularly functional biodiversity and their impacts	Garibaldi *et al*., (2017)
Agroforestry	A strategy of land management that incorporates trees or shrubs into the agricultural landscape	Leakey (2014)
Conservation tillage	A soil management approach with the aim of limited soil manipulation.	Lai (1989)
Diversified farming	Farms that integrate several crops or animals into the production system)	Garibaldi *et al*., (2017)
Ecological intensification	Emphasizes ecological processes that support production, such as nutrient cycling, biotic pest management, and pollination	Garibaldi *et al*., (2017) Kovács-Hostyánszki *et al*., (2017)
Integrated Crop Management	A whole farm approach to crop management, balancing profitability, productivity, and the environment	Lançon *et al*., (2007)
Integrated Farm Management	A whole farm approach that makes use of traditional and modern methods to increase productivity, but limit environmental impact	Leaf (2017)
Integrated Pest Management	An ecosystem approach to crop protection, in which all available measures are used to hold pest abundances below a threshold of economic damage, with an emphasis on non-chemical practices such as crop rotation, crop variety selection, hygiene, habitat management for natural enemies, biological control and monitoring	http://www.fao.org/agriculture/crops/core-themes/theme/pests/ipm/en/
Organic farming	A holistic system for enhancing soil fertility, water management, and natural control of crop pests and diseases, usually associated with low-input, small, diverse farms	Garibaldi *et al*., (2017)
Precision farming	Farming that makes use of information technology to ensure targeted and efficient management	Blackmore (1994)
Sustainable Intensification	Improving crop yield whilst improving environmental and social conditions	Garibaldi *et al*., (2017)

Source: Rosea *et al*., (2019)

1.1 Conventional and Organic Agriculture

Conventional farming, also known as industrial agriculture, refers to farming systems which include the use of synthetic chemical fertilizers, pesticides, herbicides and other continual inputs, genetically modified organisms, concentrated animal feeding operations, heavy irrigation, intensive tillage, or concentrated monoculture production. Thus conventional agriculture is typically highly resource and energy intensive, but also highly productive. Organic farming is a specific type of farming where no chemicals are added to the soil unless they are all natural. The farmers must grow their crops naturally. Organic farming is one of the several approaches found to meet the objectives of sustainable agriculture. **Organic farming is often associated directly with, *"Sustainable farming."*** However, 'organic farming' and 'sustainable farming', policy and ethics-wise are two different terms. Many techniques used in organic farming like inter-cropping, mulching and integration of crops and livestock are not alien to various agriculture systems including the traditional agriculture practiced in several countries like India. However, organic farming is a method of farming that works at grass root level preserving the reproductive and regenerative capacity of the soil, good plant nutrition, and sound soil management, produces nutritious food rich in vitality which has resistance to diseases. Organic farming is one of the widely used methods, which is thought of as the best alternative to avoid the ill effects of chemical farming. Although organic agriculture seems to be just the exclusion of synthetic external inputs but it is the ideological differences with conventional agriculture that makes organic agriculture friendly to society and environment (Table 1.2).

Table 1.2: Ideological differences between Conventional (Chemical) and Organic agriculture

Conventional Agriculture	Organic Agriculture
Reductionist approach: Targeted approach for one commodity or one pest or deficiency of nutrient- creates imbalance in system	Holistic approach: Any technology applied considering the system as a whole and No Imbalance
Domination on nature: Agriculture system is forced to produce more- regenerative capacity of natural resources decreased+ decreased productivity in long term.	Harmony with Nature: Harness the benefit of natural resources, flora and fauna by using or giving favorable environment to them- sustained productivity of natural resources
Centralize production: Produced in factories/farms, away from the place of use- no proper use of local resources+ least employment+ increase a cost of production	Decentralize production: Most of the inputs e.g. seed, manure, biopesticides etc. produced at farm/village level- suitable to local environment+ generate employment+ low cost of production
Output maximization: Over use of resources disturbs system and resources productivity in long term-Increasing cost.	Input optimization:Best use/recycling of available resources. System regenerative capacity and owners' economic capacity maintained/enhanced.
Specialization: Only one crop or tree or animal. All cost and time of nutrient and pest management has to be borne by system owner/farmer.	Diversity: Includes all possible organisms complimentary in a system. Work as mutual service providers for nutrient and pest management. Least cost and time required of system owner.

Conventional Agriculture	Organic Agriculture
Increasing input use: Target and action approach that rather, deteriorate systems regenerative capacity increasing requirement of inputs.	Decreasing input use: As the system reaching at perfection it conserve/generate its own resources e.g. for nutrition and protection- decreasing requirement of inputs
Input Intensive: Comprehensive list of chemicals with time and method. Needs experts for timely updating. Only possible in resources sufficient areas.	Knowledge Intensive: Only few resources but need how timely and best integrated. Least dependency on experts/imported technologies, once farmer trained possible in remotest area.
Cause and control approach: Most of the actions/applications are done to control the damage to system-heavy use of inputs.	Preventive, protective and proactive approach: All the actions/ applications are done in anticipation of system requirement-least use of inputs.

(*Source*: Rana, 2016)

1.2 History of Organic Farming

Traditional farming (of many particular kinds in different eras and places) was the original type of agriculture, and has been practiced for thousands of years. All traditional farming is now considered to be "organic farming" although at the time there were no known inorganic methods. For example, forest gardening, a fully organic food production system which dates from prehistoric times, is thought to be the world's oldest and most resilient agroecosystem.After the industrial revolution had introduced inorganic methods, most of which were not well developed and had serious side effects. An organic movement began in the 1940s as a reaction to agriculture's growing reliance on synthetic fertilizers and pesticides. The history of this modern revival of organic farming dates back to the first half of the 20th century at a time when there was a growing reliance on these new synthetic, non-organic methods.

1.2.1 Pre-World War II

The first 40 years of the 20th century saw simultaneous advances in biochemistry and engineering that rapidly and profoundly changed farming. The introduction of the gasoline-powered internal combustion engine ushered in the era of the tractor and made possible hundreds of mechanized farm implements. Research in plant breeding led to the commercialization of hybrid seed. And a new manufacturing process made nitrogen fertilizer — first synthesized in the mid-19th century- affordably abundant, resulted changed in the labor equation. Consciously organic agriculture (as opposed to traditional agricultural methods from before the inorganic options existed, which always employed only organic means) began more or less simultaneously in Central Europe and India. **The British botanist Sir Albert Howard is often referred to as the father of modern organic agriculture**, because he was the first to apply modern scientific knowledge and methods to traditional agriculture. From 1905 to 1924, he and his wife Gabrielle, herself a plant physiologist, worked as agricultural advisers in Pusa, West Bengal, where they documented traditional Indian farming practices and came to regard them as superior to their conventional agriculture science. Their research and further development of these methods is recorded in his writings, notably, his 1940 book**"An Agricultural Testament"** which influenced many scientists and farmers of the day.

In Germany, Rudolf Steiner's development, biodynamic agriculture, was probably the first comprehensive system of what we now call organic farming. This began with a lecture series Steiner presented at a farm in Koberwitz (Kobierzyce now in Poland) in 1924.Steiner emphasized the farmer's role in guiding and balancing the interaction of the animals, plants and soil. Healthy animals depended upon healthy plants for their food, healthy plants upon healthy soil, and healthy soil upon healthy animals for the manure. His system was based on his philosophy of anthroposophy rather than a good understanding of science.

In 1909, American agronomist F.H. King toured China, Korea, and Japan, studying traditional fertilization, tillage, and general farming practices. He published his findings in Farmers of Forty Centuries (1911, Courier Dover Publications, ISBN 0-486-43609-8). King foresaw a "world movement for the introduction of new and improved methods"of agriculture and in later years his book became an important organic reference.

The term "organic farming" was coined by Walter James (Lord Northbourne), a student of Biodynamic Agriculture, in his book Look to the Land (written in 1939, published 1940). In this text, James described a holistic, ecologically balanced approach to farming, "the farm as organism,"basing this on Steiner's agricultural principles and methods. One year previously to his book's publication, James had hosted the first biodynamic agriculture conference in England, the Betteshanger Summer School and Conference, at which Ehrenfried Pfeiffer was the key presenter.

In 1939 James, Albert Howard, Ehrenfried Pfeiffer and George Stapleton joined at Farleigh to implement an experiment comparing Biodynamic, organic and chemical fertilization methods. "The Farleigh Experiment", had been planned since initial meetings in 1936 including ten participants. The experiment was cut short due to the fact that Biodynamic compost was not available until after the Betteshanger Summer School event, the disruption of the impending war, and lack of funding. Though inconclusive, this experiment was seen as providing impetus for the similar "Haughley Experiment". In

1939 Lady Eve Balfour, who had been farming since 1920 in Haughley Green, Suffolk, England, launched the Haughley Experiment. Lady Balfour believed that mankind's health and future depended on how the soil was used, and that non-intensive farming could produce more wholesome food. The experiment was run to generate data to test these beliefs.Four years later, she published "The Living Soil" based on the initial findings of the Haughley Experiment. Widely read, it led to the formation of a key international organic advocacy group, the Soil Association.

In Japan, Masanobu Fukuoka, a microbiologist working in soil science and plant pathology, began to doubt the modern agricultural movement. In 1937, he quit his job as a research scientist, returned to his family's farm in 1938, and devoted the next 60 years to developing a radical no-till organic method for growing grain and many other crops, now known as natural farming, nature farming, or Fukuoka farming.

1.2.2 Post-World War II

Technological advances during World War II accelerated post-war innovation in all aspects of agriculture, resulting in large advances in mechanization (including large-scale irrigation), fertilization, and pesticides. In particular, two chemicals that had been produced in quantity for warfare, were reproposed for peacetime agricultural uses. Ammonium nitrate (NH_4NO_3), used in munitions, became an abundantly cheap source of nitrogen. And a range of new pesticides appeared as DDT, which had been used to control disease-carrying insects around troops, became a general insecticide, launching the era of widespread pesticide use.At the same time, increasingly powerful and sophisticated farm machinery allowed a single farmer to work larger areas of land and fields grew bigger. In 1944, an international campaign called the Green Revolution was launched in Mexico with private funding from the US. It encouraged the development of hybrid plants, chemical controls, large-scale irrigation, and heavy mechanization in agriculture around the world.

During the 1950s, sustainable agriculture was a topic of scientific interest, but research tended to concentrate on developing the new chemical approaches. One of the reasons for this, which informed and guided the ongoing Green Revolution, was the widespread belief that high global population growth, which was demonstrably occurring, would soon create worldwide food shortages unless humankind could rescue itself through ever higher agricultural technology. At the same time, however, the adverse effects of "modern farming" continued to kindle a small but growing organic movement. For example, in the US, J.I. Rodale began to popularize the term and methods of organic growing, particularly to consumers through promotion of organic gardening.

In 1962, Rachel Carson, a prominent scientist and naturalist, published Silent Spring, chronicling the effects of DDT and other pesticides on the environment.A bestseller in many countries, including the US, and widely read around the world, Silent Spring is widely considered as being a key factor in the US government's 1972 banning of DDT. The book and its author are often credited with launching the worldwide environmental movement.

In 1960, when India was facing mass shortage of food, M. S. Swaminathan along with Norman Borlaug and other scientists encouraged develop the HYV (high yielding variety) seeds of wheat and rice. This development led to Green Revolution in India and M S Swaminathan was known as 'The Father of Green Revolution'. He awarded with Padma Shri and Padma Bhushan in 1967 and 1972, respectively.

In the 1970s, global movements concerned with pollution and the environment increased their focus on organic farming. As the distinction between organic and conventional food became clearer, one goal of the organic movement was to encourage consumption of locally grown food, which was promoted through slogans like "Know Your Farmer, Know Your Food". In 1972, IFOAM (the International Federation of Organic Agriculture Movements) was founded in Versailles, France and dedicated to the diffusion and exchange of information on the principles and practices of organic agriculture of all schools and across national and linguistic boundaries.

In 1975, Fukuoka released his book, The One-Straw Revolution, with a strong impact in certain areas of the agricultural world. His approach to small-scale grain production emphasized a meticulous balance of the local farming ecosystem, and a minimum of human interference and labor.

In the U.S. during the 1970s and 1980s, J.I. Rodale and his Rodale Press (now Rodale, Inc.) led the way in getting Americans to think about the side effects of nonorganic methods, and the advantages of organic ones. The press's books offered how-to information and advice to Americans interested in trying organic gardening and farming. In 1984, Oregon Tilth established an early organic certification service in the United States.In the 1980s, around the world, farming and consumer groups began seriously pressuring for government regulation of organic production. This led to legislation and certification standards being enacted through the 1990s and to date. In the United States, the Organic Foods Production Act of 1990 tasked the USDA with developing national standards for organic products, and the final rule establishing the National Organic Program was first published in the Federal Register in 2000. In Havana, Cuba, the loss of Soviet economic support following the collapse of the Soviet Union in 1991 led to a focus on local agricultural production and the development of a unique state-supported urban organic agriculture program called organopónicos.Since the early 1990s, the retail market for organic farming in developed economies has been growing by about 20% annually due to increasing consumer demand. Concern for the quality and safety of food, and the potential for environmental damage from conventional agriculture, are apparently responsible for this trend.

1.2.3 21st Century

Throughout this history, the focus of agricultural research and the majority of publicized scientific findings has been on chemical, not organic farming. This emphasis has continued to biotechnologies like genetic engineering. One recent survey of the UK's leading government funding agency for bioscience research and training indicated 26 Genetically Modified (GM) crop projects, and only one related to organic agriculture.This imbalance is largely driven by agribusiness in general, which, through research funding and government lobbying, continues to have a predominating effect on agriculture-related science and policy.Agribusiness is also changing the rules of the organic market. The rise of organic farming was driven by small, independent producers and by consumers. In recent years, explosive organic market growth has encouraged the participation of agribusiness interests. As the volume and variety of organic products increases, the viability of the small-scale organic farm is at risk, and the meaning of organic farming as an agricultural method is ever more easily confused with the related but separate areas of organic food and organic certification.As efforts to protect our environment increase, so does the popularity of sustainable produce. Agricultural chemicals, pesticides, and fertilizers can contaminate our environment though run-off into watercourses. In the US, Certified organic standards from the EPA do not allow for the use of toxic chemicals or artificial fertilizers in organic farming.

1.3 Organic Cycle Optimization

Organic farming is considered a promising solution for reducing environmental burdens related to intensive agricultural management practices. These changes in agricultural practices led to numerous environmental problems like high consumption of non-renewable energy resources, loss of biodiversity, pollution of the aquatic environment by the nutrients nitrogen and phosphorus as well as by pesticide (Fig. 1.1). Each field, farm, or region contains a given quantity of nutrients. Management should be used in such a way that optimal use is made of this finite amount.

- This means that the nutrients should be recycled and used a number of times in different forms.
- Care should be taken that only a minimum amount of nutrients actually leave the system so that import of nutrients can be restricted.

The quantity of nutrients available to plants and animals can be increased within the system by activating the edaphon, resulting in increased weathering of parent material.

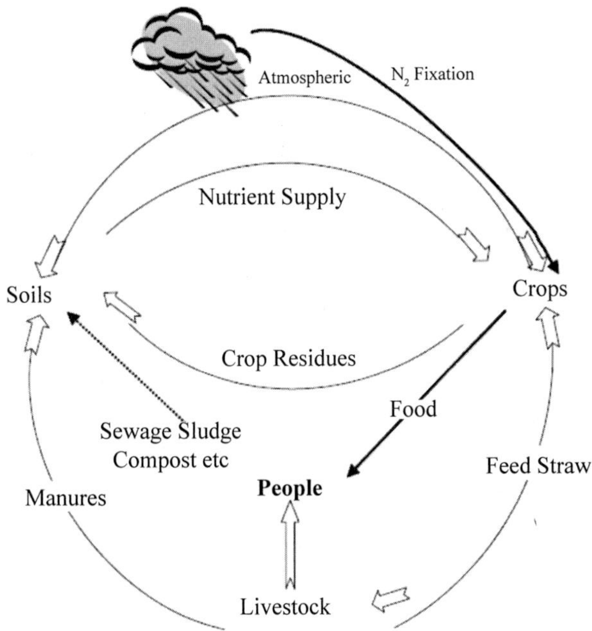

Fig. 1.1: Organic Cycle of an Organic Farming System

1.4 Milestones of "The World of Organic Agriculture"

Year Milestone

1999 The organizers of the Biofach organic trade fair ask the German Foundation Ecology & Agriculture (SÖL) to compile basic data on organic agriculture worldwide. 11Mha of organic farmland; 0.2 Million producers, 15.2 billion US dollars of retail sales.

2000 The first edition of the yearbook with global organic farming statistics ("Organic Agriculture Worldwide") is published by SÖL in collaboration with Biofach and IFOAM – Organics International and presented at the Biofach Fair, then in Frankfurt, Germany. Since then, the data have been published annually and are presented at Biofach every year.

2001 The Research Institute of Organic Agriculture FiBL joins as a partner.

2006 For the first time, land use and crop data on organic agriculture are collected.

2007 The World of Organic Agriculture, 2007 edition, is translated into Chinese.

2008 Funding by the Swiss State Secretariat of Economic Affairs (SECO) and the International Trade Centre (ITC) enables FiBL to set up a professional database to improve data collection, processing, storage, and analysis. FiBL sets up the Organic-World.net website.34.5 Mha of organic farmland, 1.4 Million producers, 50.2 Billion US dollars in retail sales (data published in 2010).

2011 The Food and Agriculture Organisation of the United Nation (FAO) includes the FiBL data into its FAOSTAT online database.

2012 The Organic Data Network project starts, funded under the 7th Framework Progamme for Research and Technological Development of the European Union. Under this project, the data collection andprocessing methods are improved and an interactive online database is set up. Organic and Beyond (China) translates and publishes an excerpt of "The World of Organic Agriculture" (annually since that year).

2013 The "The World of Organic Agriculture" is translated into Korean (also the 2014 edition).

2014 The follow-up project of the SECO-ITC-funded project "Global Information System for Organic Market and Production Data" includes data collection on Voluntary Sustainability Standards.

2015 The first edition of "The State of Sustainable Markets" with data on 14 Voluntary Sustainability Standards is published by FiBL, ITC and the International Institute of Sustainable Development (IISD).

2017 69.8Mha of organic farmland, 2.9 Million producers, 97 Billion US dollars in retail sales (data published in 2019).

2018 FiBL launches a dedicated website for the interactive online database – Statistics.FiBL.org.

2019 20th edition of "The World of Organic Agriculture" is launched at Biofach.Funding by the Sustainability Fund of Coop Switzerland (Coop Fonds für Nachhaltigkeit). The 2018 edition of "The World of Organic Agriculture" is translated into Persian.

(*Source*: FiBL & IFOAM (2019)

1.5 Milestones of the Global Organic Market

Year Milestone

1999 Global organic market data published for the first time by the International Trade Centre ITC, Geneva (Organic food and beverages: World supply and major European markets).

2001 Organic Monitor (now known as Ecovia Intelligence) formed to analyze the global market for organic products. Global organic food sales reach 20 billion US dollars. The Bio-Siegel national organic logo is launched in Germany.

2002 Japanese Agricultural Standard for organic foods is introduced.

2002 USDA National Organic Programme (NOP) is implemented.

2004 EU expansion leads to 14% rise in European organic farmland.Korean government introduces national organic programme.

2005 China introduces national organic standard.

2008 Global organic food sales surpass USD 50 billion.

2009 EU implements revised organic regulations.Brazil introduces national organic standard.US enters organic equivalency agreement with Canada. Europe temporarily overtakes North America to have largest organic food market.

2010 Introduction of the current EU organic logo.

2012 New US-EU organic equivalency agreement.

2013 Whitewave Foods acquires Earthbound Farm and becomes major organic food enterprise.

2014 Surge in demand for organic baby food gives China largest organic infant formula market.

2016 US passes GM labeling bill, whilst Non-GMO project product sales reach 20 billion US dollars.

2017 Amazon buys the world's leading natural & organic food retailer Whole Foods Market.

(*Source*: FiBL & IFOAM 2019)

1.6 Milestones of IFOAM – Organics International

Year Milestone

1972 IFOAM - The International Federation of Organic Agriculture Movements – is founded in Versailles, France,registered as FIMAB (Fédération Internationale des Mouvements d'Agriculture); the office is hosted by Nature et Progrès.

1973 First "Circular letter" of IFOAM comes out. By the end of 1976, 19 newsletters are published.

1974 At the second General Assembly of IFOAM, members meet at the Nature et Progrès congress in Paris.

1975 Nature et Progrès based in Sainte-Geneviève-des-Bois, just south of Paris, remains responsible for the secretariat, which is itself situated in La Feuillée, Brittany, France.

1976 A third General Assembly of IFOAM members takes place in Seengen, Switzerland (80 members), where it is decided to define organic farming with international standards and use only the English language.

1977 The IFOAM office is moved to Oberwil near Basel in Switzerland, hosted by the Research Institute of Organic Agriculture FiBL.

1977 1st IFOAM International Scientific Conference "Towards a Sustainable Agriculture" takes place in Sissach, Switzerland.

1978 IFOAM moves its office to Topsfield, Massachusetts, USA.

1980 The first version of the "Recommendations for international standards of biological agriculture" is accepted by the biennial IFOAM General Assembly in Brussels.

1982 The IFOAM recommendations become the "Standards of biological agriculture for international trade and national standards", with validity restricted to two years.

1982 The head office moved to Grasse, in the South of France. Thanks to better financial support, a professional secretariat is funded from 1987.

1984 IFOAM membership exceeds 100 member organisations.

1986 IFOAM receives observer status at the UN department of information.

1986 The European Commission starts its work on the regulation of organic food production, and IFOAM is the logical counterpart.

1987 IFOAM hires its first Executive Director, and the office is moved to Tholey-Theley, Germany.

1990 The first Biofach organic trade fair is held, and the long-lasting partnership between IFOAM and Biofach begins.

1991 Council Regulation (EEC) 2092/91 on organic production is published, coming into force in 1993, which was welcomed by IFOAM.The Codex Alimentarius Commission, a joint FAO/WHO Food Standards Programme (with strong involvement of IFOAM as observer organizations), begins developing guidelines for the production, processing, labelling and marketing of organically produced foods.

1992 IFOAM influences the UN Conference on Environment and Development in Rio de Janeiro.

1996 FAO organizes the first World Food Summit, with IFOAM participation; one year later, IFOAM is officially recognized as the global representative of the organic movement.

1997 IFOAM founds the International Organic Accreditation Service (IOAS).

1999 Codex Alimentarius approves the first plant production guidelines for organically produced food,significantly influenced by the IFOAM basic standards and Council Regulation (EEC) 2092/91 on organic production.

2001 Codex Alimentarius approves the first animal production guidelines for organically produced food,influenced by IFOAM basic standards and the EU regulation for organic livestock 1804/99.

2002 International IFOAM conference on Organic Guarantee Systems with FAO and UNCTAD.

2003 IFOAM moves offices to Bonn, the German UN City.

2005 The General Assembly adopts the principles of organic agriculture: health, ecology, fairness and care.

2007 IFOAM successfully challenges a patent on a fungicide derived from the seed of the neem tree.

2008 The General Assembly ratifies the definition of organic agriculture.

2016 The Sustainable Organic Agriculture Action Network (SOAAN) produces the Best Practice Guidelines, and the global organic movement discusses Organic 3.0.

(*Source*: FiBL & IFOAM 2019)

1.7 Definition and Options in Organic Farming

The concept of organic farming is not new to India as it has existed here before the introduction of modern methods of farming. It is actually based on the minimal use of off farm resources (inputs) and on management practices that restore, maintain and enhance ecological harmony. The international Federation of organic Agriculture Movements (IFOAM) has defined it as Production system that sustains the health of soils, ecosystems and people. It relies on ecological processes, biodiversity and cycles adapted to local conditions, rather than the sum of inputs with adverse effects. According to the Agricultural and Processed Food Products Export Development Authority (APEDA) of India "Organic products are grown under a system of agriculture without use of chemical fertilizers and pesticides with an environmentally and socially responsible approach". This is a method of farming that works at grass root level preserving the reproductive and regenerative capacity of the soil, good plant nutrition, and sound soil management produces nutritious food rich in vitality which has resistance to diseases. As per the definition of the United States Department of Agriculture (USDA) study team on organic farming "Organic farming is a system which avoids or largely excludes the use of synthetic inputs (such as fertilizers, pesticides, hormones, feed additives etc.) and to the maximum extent feasible rely upon crop rotations, crop residues, animal manures, off-farm organic waste, mineral grade rock additives and biological system of nutrient mobilization and plant protection". FAO suggested that "Organic agriculture is a unique production management system which promotes and enhances agro-ecosystem health, including biodiversity, biological cycles and soil biological activity, and this is accomplished by using on-farm agronomic, biological and mechanical methods in exclusion of all synthetic off-farm inputs".

There are three options in organic farming as-

a) **Integrated farming system:** Recycling of local organic resources generated from dairying, fisheries, poultry, goat farming, mushroom production, agro-forestry and agriculture system. This is log input organic farming.

b) **Integrated conventional and organic farming technology**: This involves integrated plant nutrients supply (IPNS) and integrated pest management (IPM). Bio-fertilizers could serve as alternate sources of nitrogen or phosphorus or both.

c) **Pure organic farming:** Use of organic manures/compost, recycling of plant residues, bio-fertilizers and bio-pesticides.

1.8 Key Characteristics of Organic Farming

- Providing crop nutrients indirectly using relatively insoluble nutrient sources which are made available to the plant by the action of soil micro-organisms.
- Nitrogen self-sufficiency through the use of legumes and biological nitrogen fixation, as well as effective recycling of organic materials including crop residues and livestock manures.
- Careful attention to the impact of the farming system on the wider environment and the conservation of wildlife and natural habitats.
- Weed, disease and pest control relying primarily on crop rotations, natural predators, diversity, organic manuring, resistant varieties and limited (preferably minimal) thermal, biological and chemical intervention.
- Protecting the long term fertility of soils by maintaining organic matter levels, encouraging soil biological activity, and careful mechanical intervention.
- The extensive management of livestock, paying full regard to their evolutionary adaptations, behavioral needs and animal welfare issues with respect to nutrition, housing, health, breeding and rearing.

1.9 Need of Organic Farming

With the increase in population our compulsion would be not only to stabilize agricultural production but to increase it further in sustainable manner. The scientists have realized that the 'Green Revolution' with high input use has reached a plateau and is now sustained with diminishing return of falling dividends. Thus, a natural balance needs to be maintained at all cost for existence of life and property. The obvious choice for that would be more relevant in the present era, when these agrochemicals which are produced from fossil fuel and are not renewable and are diminishing in availability. It may also cost heavily on our foreign exchange in future. Modern, intensive agriculture causes many problems, including the following:

a) Artificial chemicals destroy soil micro-organisms resulting in poor soil structure and aeration and decreasing nutrient availability.

b) Artificial fertilizers and herbicides are easily washed from the soil and pollute rivers, lakes and water sourses.

c) Artificial pesticides can stay in the soil for a long time and enter the food chain where they build up in the bodies of animals and humans, causing health problems.

d) The prolonged use of artificial fertilizers results in soils with a low organic matter content which is easily eroded by wind and rain.

e) Pests and diseases become more difficult to control as they become resistant to artificial pesticides. The number of natural enemies decrease because of pesticide use and habitat loss.

f) Dependency on fertilizers, greater amounts are needed every year to produce the same yields of crops.

1.10 Principles of Organic Farming

The four principles of organic agriculture are as follow (Fig. 1.2) as:

1.10.1 Principle of Health

Organic Agriculture should sustain and enhance the health of soil, plant, animal, human and planet as one and indivisible. This principle points out that the health of individuals and communities cannot be separated from the health of ecosystems - healthy soils produce healthy crops that foster the health of animals and people. Health is the wholeness and integrity of living systems. It is not simply the absence of illness, but the maintenance of physical, mental, social and ecological well-being. Immunity, resilience and regeneration are key characteristics of health. The role of organic agriculture, whether in farming, processing, distribution, or consumption, is to sustain and enhance the health of ecosystems and organisms from thesmallest in the soil to human beings. In particular, organic agriculture is intended to produce high quality, nutritious food that contributes to preventive health care and well-being. In view of this it should avoid the use of fertilizers, pesticides, animal drugs and food additives that may have adverse health effects.

1.10.2 Principle of Ecology

Organic Agriculture should be based on living ecological systems and cycles, work with them, emulate them and help sustain them. This principle roots organic agriculture within living ecological systems. It states that production is to be based on ecological processes, and recycling. Nourishment and well-being are achieved through the ecology of the specific production environment. For example, in the case of crops this is the living soil; for animals it is the farm ecosystem; for fish and marine organisms, the aquatic environment.Organic farming, pastoral and wild harvest systems should fit the cycles and ecological balances in nature. These cycles are universal but their operation is site-specific. Organic management must be adapted to local conditions, ecology, culture and scale. Inputs should be reduced by reuse, recycling and efficient management of materials and energy in order to maintain and improve environmental quality and conserve resources. Organic agriculture should attain ecological balance through the design of farming systems, establishment of habitats and maintenance of genetic and agricultural diversity. Those who produce, process, trade, or consume organic products should protect and benefit the common environment including landscapes, climate, habitats, biodiversity, air and water.

1.10.3 Principle of Fairness

Organic Agriculture should build on relationships that ensure fairness with regard to the common environment and life opportunities. Fairness is characterized by equity, respect, justice and stewardship of the shared world, both among people and in their relations to other living beings. This principle emphasizes that those involved in organic agriculture should conduct human relationships in a manner that ensures fairness at all levels and to all partners-farmers, workers, processors, distributors, traders and consumers. Organic agriculture should provide everyone involved with a good quality of life, and contribute to food sovereignity and reduction of poverty. It aims to produce a sufficient supply of good quality food and other products. This principle insists that animals should be provided with the conditions and opportunities of life that accord with their physiology, natural behavior and well-being. Natural and environmental resources that are used for production and consumption should be managed in a way that is socially and ecologically just and should be held in trust for future generations. Fairness requires systems of production, distribution and trade that are open and equitable and account for real environmental and social costs.

1.10.4 Principle of Care

Organic Agriculture should be managed in a precautionary and responsible manner to protect the health and well-being of current and future generations and the environment. Organic agriculture is a living and dynamic system that responds to internal and external demands and conditions. Practitioners of organic agriculture can enhance efficiency and increase productivity, but this should not be at the risk of jeopardizing health and well-being. Consequently, new technologies need to be assessed and existing methods to be reviewed. Given the incomplete understanding of ecosystems and agriculture, care must be taken. This principle states that precaution and responsibility are the key concerns in management, development and technology choices in organic agriculture. Science is necessary to ensure that organic agriculture is healthy, safe and ecologically sound. However, scientific knowledge alone is not sufficient. Practical experience, accumulated wisdom and traditional and indigenous knowledge offer valid solutions, tested by time. Organic agriculture should prevent significant risks by adopting appropriate technologies and rejecting unpredictable ones, such as genetic engineering. Decisions should reflect the values and needs of all who might be affected, through transparent and participatory processes.

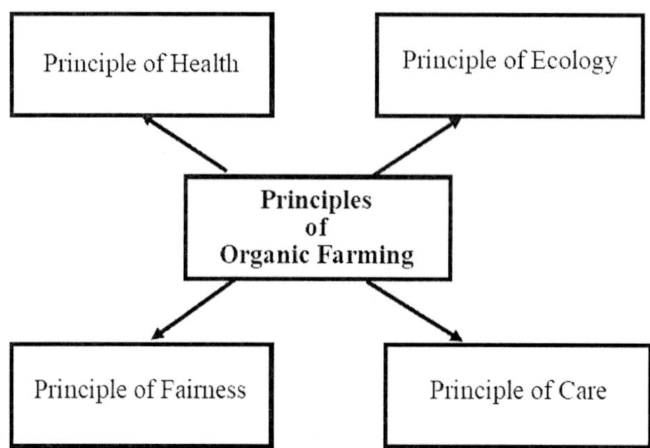

Fig. 1.2: Principles of Organic Framing

1.11 Current Global and Indian Scenario

In 2017, 69.8 Mhal and were under organic agricultural management worldwide.The region with the most organic agricultural land is Oceania, with 35.9 Mha, followed by Europe with 14.6 Mha, Latin America (8 Mha), Asia (6.1 Mha), North America (3.2 Mha), and Africa (2.1 Mha). Oceania has half of the global organic agricultural land. Europe, a region that has had a very constant growth of organic land over the years, has over 20% of the world's organic agricultural land followed by Latin America with 11%. Australia, which continued to experience growth of its organic area in 2017 (+8.5 Mha), is the country with the most organic agricultural land; it is estimated that 97% of the farmland are extensive grazing areas. Argentina is second followed by the United States in third place (Table 1.3a). The 10 countries with the largest organic agricultural areas have a combined total of nearly 55 Mha and constitute three-quarters of the world's organic agricultural land.The share of the world's agricultural land that is organic is 1.4%.The highest organic share of total agricultural land, by region, is in Oceania (8.5%) followed by Europe with 2.9% and Latin America with 1.1%. In the European Union, the organic share of the total agricultural land is 7.2%. In the other regions, the share is less than one percent (Table 1.3b). Many individual countries, however, have a much higher organic share, and in 14 countries, 10% or more of the agricultural land is used for organic production. Most of these countries are in Europe. The countries with the highest organic share are Liechtenstein and Samoa, with almost 38 percent of its agricultural land under organic management. It is interesting to note that many island states have high shares of agricultural land under organic management, such as Samoa and Sao Tome and Principe. However, 56% of the countries for which data is available have less than one percent of their agricultural land under organic management (Table 1.4).

Table 1.3: World: Organic agricultural land (including in-conversion areas) and regions' shares of the global organic agricultural land 2017 and organic share of total agricultural land by region 2017

		A	B
Regions'	Organic agricultural land [hectares]	Regions shares of the global organic agricultural land	Share of total agri. land
Africa	2'056'571	3%	0.2%
Asia	6'116'834	9%	0.4%
Europe	14'558'246	21%	2.9%
Latin America	8'000'888	11%	1.1%
North America	3'223'057	5%	0.8%
Oceania	35'894'365	51%	8.5%
World	69'845'243	100%	1.4%

(*Source*: FiBL & IFOAM (2019)

Table 1.4: Organic Agriculture: Key Indicators and Top Countries (2017)

Indicator	World	Top countries
Country with organic activities	181 countries	
Organic agricultural land	69.8Mha (1999: 11Mha)	Australia (35.6Mha) Argentina (3.4Mha) China (3.0Mha)
Organic share of total agricultural land	1.4 %	Liechtenstein (37.9 %) Samoa (37.6 %) Austria (24.0 %)
Wild collection and further non-agricultural areas	42.4Mha (1999: 4.1Mha)	Finland (11.6Mha) Zambia (6.0Mha) Tanzania (2.4Mha)
Producers	2.9 million producers (1999: 200'000 producers)	India (835'000) Uganda (210'352) Mexico (210'000)
Organic market	97 billion US dollars*2 (approx. 90 billion euros) (2000: 17.9 billion US dollars)	US (45.2 billion US dollars; 40 billion euros) Germany (11.3 billion US dollars; 10 billion euros) France (8.9 billion US dollars; 7.9 billion euros)

Indicator	World	Top countries
Per capita consumption	12.8 US dollars (10.8 euros)	Switzerland (325 US dollars; 288 euros) Denmark (315 US dollars; 278 euros) Sweden (268 US dollars; 237 euros)
Number of countries with organic regulations	93 countries Number of affiliates of IFOAM – Organics International 2018: 726 affiliates from 110 countries	Germany - 76 affiliates India - 47 affiliates China - 45 affiliates United States - 43 affiliates

(*Source*: FiBL & IFOAM 2019)

India has a remarkable potential to produce all varieties of organic products, owing to the existence of various agroclimatic zones within its borders. The total area under organic certification was 5.71 Mha in 2015-16 (Fig 1.3). This included 26% cultivable area with 1.49 Mha and 74% (4.22 Mha) forest and wild area for collection of minor forest produce.The organic production area in India falls essentially undertwo management systems i.e. (i) National Programme on Organic Production (NPOP) and (ii) Participatory Guarantee System-India (PGS-India).

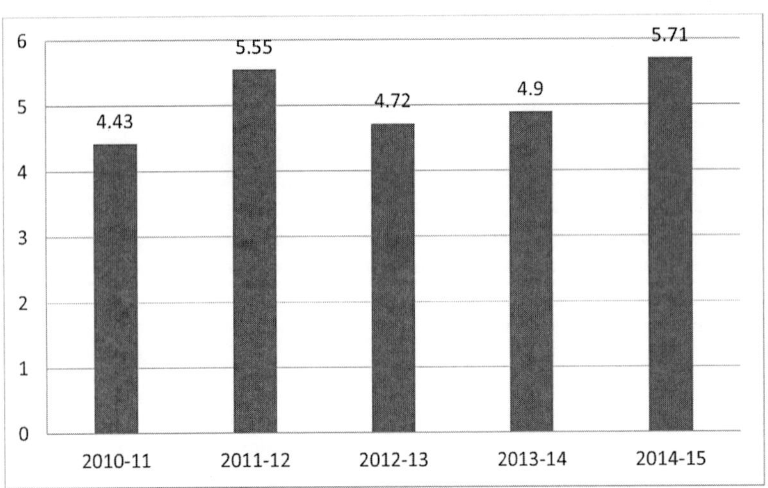

Fig 1.3: Organic area in India (in million ha)

Source: ASSOCHAM, 2018

There was an increase in area at a CAGR of 6% from 2010-11 to 2015-16 and absolute growth of 29% during the same period. It is likely to grow at a rate of 8-10% till 2020. Among the states, Madhya Pradesh has the largest area under organic certification (4.62 lakh ha) followed by, Maharashtra (1.98 lakh ha) Rajasthan (1.55 lakh ha), Telangana (1.04 lakh ha), Odisha (0.96 lakh ha), Karnataka (0.94 lakh ha), Gujarat (0.77 lakh ha) and Sikkim (0.76 lakh ha) (ASSOCHAM, 2018). These states had a combined share of 90% of the area under organic certification in 2015-16 (Fig 1.4, 1.5).More than 157,721 organic cotton farmers were certified to the NPOP Standard in 2014-15.In the same year, 276,736 ha of land was certified as organic with approximately 148,105 ha (54%) dedicated to organic cotton (Fig 1.6).Prime Minister Sh. Narendra Modi addressed the 95th Annual Plenary Session of the Indian Chamber of Commerce (ICC) through virtual mode. He said the North East Indian region can become a hub of organic farming and the capital for organic products. He said, Just like Sikkim, the entire North-East can become a huge hub for organic farming (Krishna, 2020). The government has been promoting organic farming in the country through various central schemes. The move is aimed at reducing the use of chemical fertilisers, pesticides and growth regulators. States such as Madhya Pradesh, Gujarat, Maharashtra and Sikkim have boosted organic farming by providing various incentives and support. In a major push to organic farming in the country, the Agriculture Ministry has proposed to double the allocation for the sector to Rs 1,300 crore annually. The Centre has provided Rs. 660 crore to support the sector in its budget estimate (BE) for FY21. In a presentation to the Fifteenth Finance Commission, the Agriculture Ministry has proposed to bring additional 25 lakh hectares under organic farming in the next 5 years. The current organic farming coverage of 28 lakh hectares is a mearly 2% of the total farm land (Kumar, 2020).

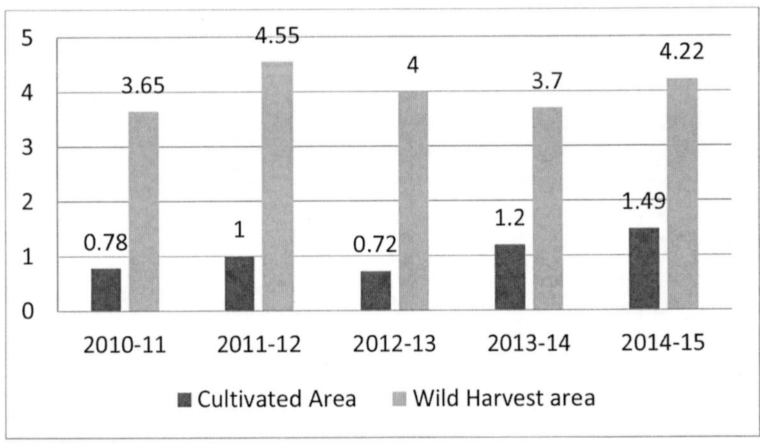

Fig 1.4: India's cultivated and wild harvest area (in million ha)

Source: ASSOCHAM, 2018

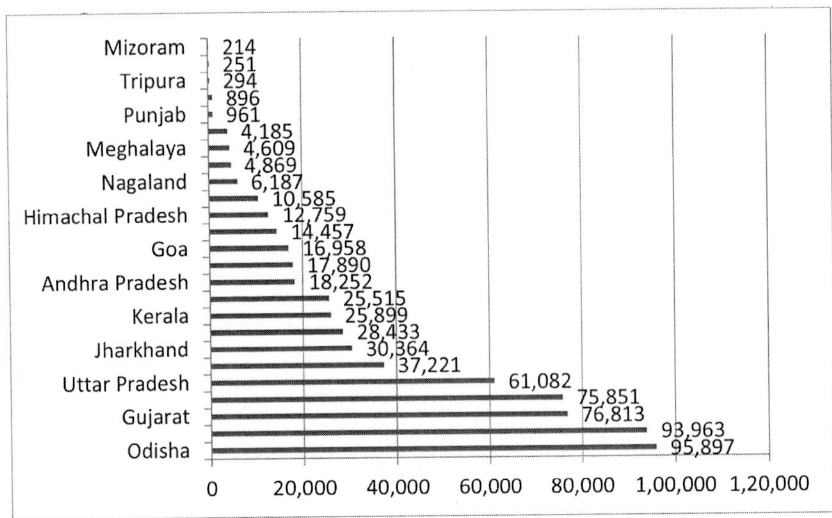

Fig 1.5: State-wise area (inha) of Organic Agriculture for 2014-2015

Source: ASSOCHAM, 2018

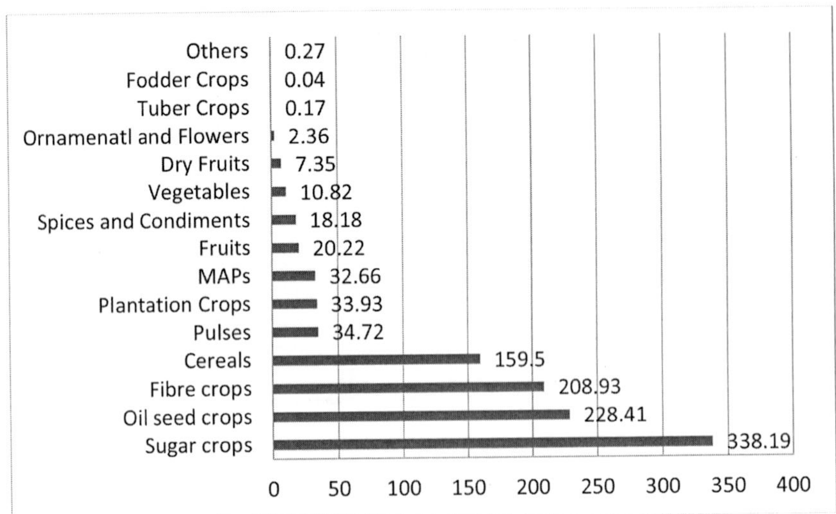

Fig. 1.6: Production of Organic Crops in India by category in thousand MT for 2014-15

Source: ASSOCHAM, 2018

According to the IFPAM 2019, the total area dedicated to organic agriculture in Asia was almost 6.1 Mha in 2017. There were 1.1 million producers, most of which were in India. The leading countries by area were China (3 Mha) and India (almost 1.8 Mha); Timor-Leste had the highest proportion of organic agricultural land (8.7%). Twenty-two countries in the region have legislation on organic agriculture, and six countries are in the process of drafting legislation. Nine countries have a national standard but no organic legislation (Fig 1.7, 1.8).

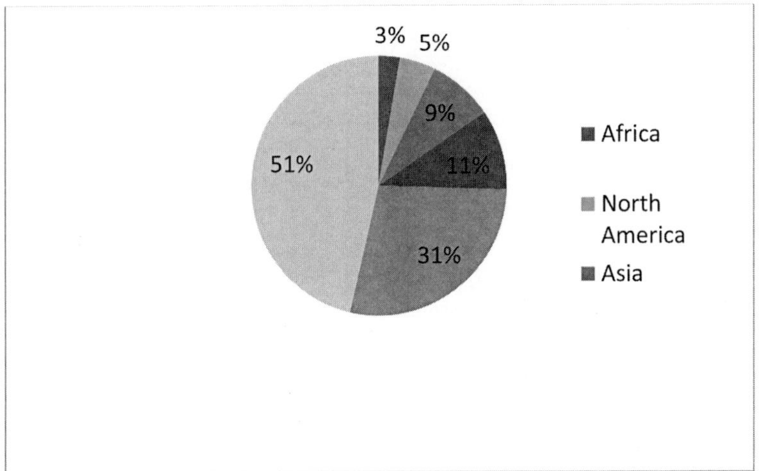

Fig 1.7: World: Distribution of organic agricultural land by region 2017
Source: FiBL Survey 2019

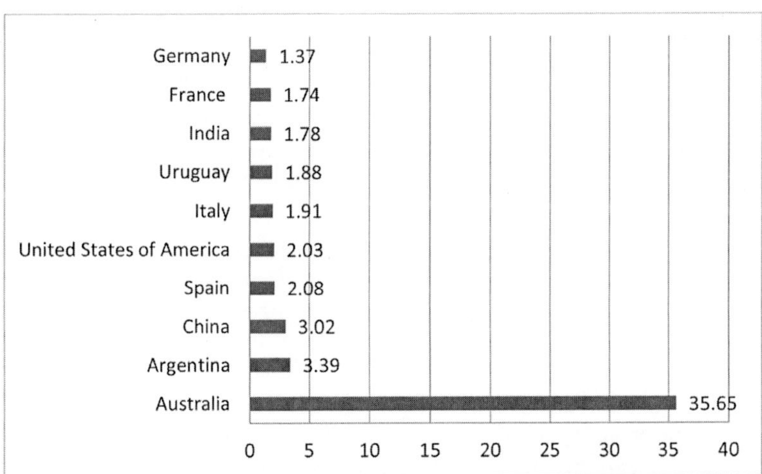

Fig. 1.8: World: The ten countries with the largest areas of organic agricultural land 2017
Source: FiBL Survey 2019

Organic farming deliver the majority of the environmental benefits (Fig. 1.9)

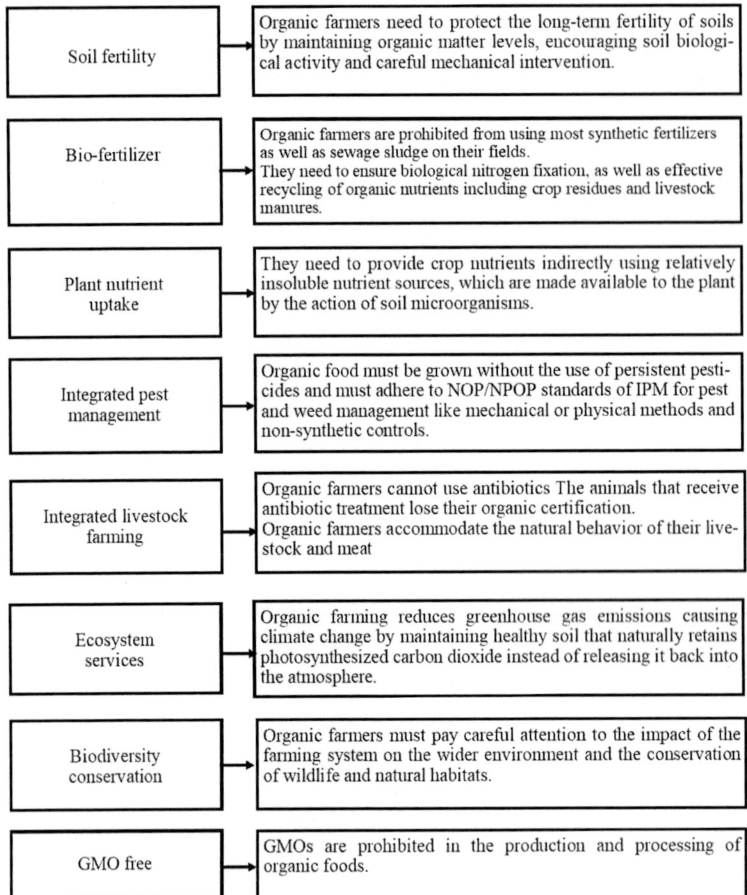

Fig 1.9: Environmental benefits of Organic Farming

1.12 Trend of Organic Food Consumption in India

The marketing of food is a matter of consideration as the production of food is determined by the vagaries of temperature as well as the various marketing constraints at national level as well as international level. The scope for marketing organic food in India is vast and still yet unexplored to its full potential. India experienced good growth in the organic business sector, with exports reportedly growing between 25 and 30% and domestic markets growing at about 40% (Hill, 2016). Two initiatives were launched by India's central government: allocation of 100 crore rupees for organic market development in the Northeast and launch of the government's Participatory Guarantee Systems (PGS) programme with a pledge of 300 crore rupees for 2015-16 (OFAI, 2016). Domestic market is also growing at an annual growth rate of 15-25%. India produced around 1.35 million MT (2015-16) of certified organic products which includes all varieties of food products namely sugarcane, oil seeds,cereals & millets, cotton, pulses, medicinal plants, tea,fruits, spices, dry fruits, vegetables, coffee, cotton, fibre etc. (Yadav, 2015). Though 50% of the organicfood production in India is targeted towards exports, there are many who look towards organic food fordomestic consumption (Rekha and Neeraj, 2013) andthe increasing trend in organic food consumption is evident from the fact that many organic food stores are mushrooming in India.

India's GDP growth of 6.5% in 2017 was strong despite challenges like the implementation of GST. The forecast for GDP growth in FY2019 is predicted to be more than 7%. This will certainly improve the performance of different industries like Organic Foods, Pharmaceuticals and FMCG.According to TechSci Research report, Global organic food market stood at $110.25 billion in 2016 and is projected to grow at a CAGR of 16.15%, in value terms, during 2017 to 2022, to reach $262.85 billion by 2022.With Indian Organic Food industry growing in double-digit during 2013-2017, it would not be wrong to say that the industry will perform well in 2019. There are many factors that have contributed to this growth until now.

- Firstly, growth in e-commerce sector has acted like a facilitator for the organic food industry to reach out to the potential consumers in Tier II and Tier III cities.
- Secondly, with free/low-cost access to the Internet, more and more people are getting to learn about the benefits of organic food. Therefore, the demand has gone up during 2018.
- Lastly, the industry landscape is becoming competitive with more and more players entering the industry. Going by the trend in 2018, the organic food industry in 2019 will grow at a faster pace. Here are a few factors that will accelerate the growth of the organic food industry:

1.13 The Rise in Demand for Organic Food

As the digital literacy rate in India is growing, the demand for organic food in India will see good growth. People are becoming aware of the harmful effects of chemicals and fertilizers used during farming in India. There is a great rise in awareness of organic farming methods, making the food more healthy and nutritious.

1.1.1 Higher Spending Capacity of the Middle Class: According to a research by the India Brand Equity Foundation, the spending capacity of the

Indian middle class is anticipated to double by 2025. With the rise in disposable income and access to credit, the Indian middle class will have a high spending capacity on organic food.

1.1.2 Government Regulations: The Government of India is promoting Organic Farming in the country. Hon'ble Prime Minister, Mr.Narendra Modi introduced Paramparagat Krishi VikasYojana (PKVY), under which 2 lakh hectares has been made suitable for organic farming thereby benefitting 5 lakh farmers. This step will further boost organic farming in India in 2019 onwards.

1.1.3 The Rise in the Export of Indian Organic Food: According to the Agricultural and Processed Food Products Export Development Authority (APEDA), the demand for Indian organic food products is on the constant increase worldwide as India exported organic products worth $515 million in the financial year 2017-18, from $370 million in 2016-17. This trend will further accelerate in 2019.

1.1.4 Supply Chain Enhancement: One of the major issues in the organic food market is the mismanagement in the supply chain of organic fruits and vegetables. The main challenge is related to the cost and logistics which are included in the moving of locally or regionally produced organic produce. In the case of organic supplies, the time of reaching the fruits and vegetables from the time of the production to the end consumers matters the most due to its less shelf life. But since the demand of the organic products is rising, the need for solving the problems in the supply chain management is going to arise and the enhancement in the supply chain is expected in 2019.

1.1.5 Traditional and Community-based Management: In some communities, the adoption of conventional agriculture has substituted traditional cultivation systems with high biodiversity for monocrops of genetically similar individuals. In a relatively short period of time, such systems have led to environmental degradation, social disintegration and misery within communities. However, many traditional agricultural systems that have been the basis of food security and community cultures have since been saved through organic agriculture approaches. The introduction of organic management, based on traditional experiences and new knowledge of natural processes, has allowed the maintenance of the agro-ecological systems and has improved socio-economic conditions of rural communities, especially in environmentally vulnerable areas. These agricultural systems are also based on strong farmer participation in the decision-making process, exchange of information and distribution of benefits. Some of the examples illustrate the community-based rehabilitation of abandoned and degraded agro-ecosystems, through organic agriculture, in flood plains of Bangladesh and mountainous areas of Indonesia and Mexico.

1.13.5.1 Nayakrishi Andolon - A Community-based System of Organic Farming, Bangladesh

In the flood plains of Bangladesh, community-based organic agriculture resulted from an increasing awareness to the harmful effects of the Green Revolution. The latter was showing a tremendous decline in crop yields despite an enormous increase in the need for the application of fertilizers and pesticides. Groundwater was less available, livestock and fish populations were diminishing, the health situation was worsening (including gastric, skin and respiratory diseases) and exogenous varieties were gradually replacing traditional varieties. This forced many poor farmers to sell their land and other productive assets, shifting from farming to non-farming occupations. Following particularly terrible floods in 1988, some farmers, together with UBINIG (Policy Research for Development Alternatives), a non-governmental organization, gathered together to seek an alternative - not just an alternative method of farming, but community-based work, which is organic in nature. They named it Nayakrishi Andolon (New Agriculture Movement). Initially, the peasant women took the lead in stopping the use of pesticides, mainly for health reasons. Then, a group of farmers organized themselves to experiment with green manure and compost. Compost made of water hyacinth, available in plenty, became quite popular and soon Nayakrishi Andolon spread from village to village. As experience and confidence grew, the farmers developed a set of ten simple principles for Nayakrishi farming, all focusing on the use of locally available resources to enhance the efficiency of land, water, biodiversity and energy as well as the control over seed within the farming community.

In Nayakrishi villages, farmers derive more varieties of fish, together with a wide range of uncultivated crops, which either come as accompanying crops due to multiple cropping in the fields, or grow on the common land as no more herbicides are used. Livestock and poultry also develop more rapidly, thereby enriching the food security of the people. Similarly, the planting of local-variety trees is an integral part of the practice in Nayakrishi villages, which, in turn, attracts birds, butterflies and other pollinators and predators. Every village where Nayakrishi is actively adopted has its own *gram karmi* (extension workers). Apart from networking and campaigning for Nayakrishi, *gram karmi* maintain audits of the natural resources of the village. This information is pooled collectively and is a vital practice in maintaining and managing the local biodiversity. The Nayakrishi farmers can easily be put on alert if it appears that any land race or wild species or variety is disappearing or being lost.Around 65,000 families from all over Bangladesh now follow Nayakrishi principles and the movement is spreading fast. Most important is the general confidence among farmers that Nayakrishi is "economically viable", but the ecological situation is also improving, the land is regaining fertility and biodiversity is being strengthened.

1.13.5.2 Ladang Cultivation of Organic Spices in Sumatra, Indonesia

On the island of Sumatra, Indonesia, ruthless exploitation by forestry, fishing and extractive industries in the last decades has decreased the rich biodiversity of the region. Despite this, some areas still survive in a state close to that of pre-European colonization, mainly because of their mountainous, remote locations. Some of these areas form part of national parks that still support the rare Sumatra tiger, rhinoceros and elephant, as well as some native people following traditional lifestyles in the forest.Many poor farmers live around these areas and use slash-and-burn techniques for the production of crops for self-sufficiency and for the market. These practices threaten the remaining forest areas and even the national parks on whose boundaries they encroach.

Thomas Fricke, a former advisor to the United Nations, World Wildlife Fund and Indonesian national parks on sustainable agriculture projects, aimed to find a solution to that problem. In 1995, he and his wife Sylvia Blanchet founded ForesTrade, an international company dedicated to preserving biodiversity through responsible trade. In 1996 ForesTrade, in collaboration with local NGOs and the National Parks of Indonesia began the Indonesian Cassia Cinnamon Project, encouraging local farmers to stop clear-cutting the rainforest. The project focused on land bordering a national forest park, providing a buffer zone for the protection of biodiversity in the rapidly disappearing forested areas. Some of the local people joined forces with ForesTrade, creating grower groups for the production of organic spices for the European and the United States markets. Good prices, generally a little better than could be expected from the conventional market, are paid to the farmers, together with a bonus to the local community. The community bonus is used to run training centres, nurseries and other community services. These farmers agree to follow organic agriculture practices, avoiding the use of chemical pesticides and fertilizers. Certified-organic growers in Sumatra produce a variety of spices and essential oil crops, such as chilli, turmeric, ginger, vanilla, cloves, cardamom, nutmeg (and mace), black and white pepper, patchouli and cassia (cinnamon). They are inspected and certified by the National Association for Sustainable Agriculture (Australia), by the Dutch certification body SKAL or by Oregon Tilth. In a short time approximately, 3000 Sumatran farmers have begun producing organic spices for the world markets. This has led to improved socio-economic conditions for communities while at the same time preserving biodiversity both in the national parks and in the local agroforestry systems (garden/forest plots).

1.13.5.3 Organic Coffee Production and Biodiversity Management, Chiapas, Mexico

At the end of 1980s, small coffee producers (most of them Tzotzil and Tzeltal indigenous people) in San Cristóbal de Las Casas, the highlands of Chiapas, southern Mexico, faced a deep economic crisis due to the fall in coffee prices on the world market. Together with the disappearance of direct support for coffee growers in terms of technical assistance, marketing and financial support, this resulted in the abandonment of practices for maintaining the crop and processing the beans, leading to lower yields and product quality. Many of these farmers have organized themselves in 1983 into the Beneficio "Majomut" Coffee Growers' Union of Ejidos and Communities, a grassroots social organization with 1450 membersin 25 communities. The organization was created as a means to bring together farmers in the processing and direct marketing of their coffee. Members work on average of two hectares and cultivate corn, beans and coffee. As coffee is sold, it forms the main source of family income. Gradually, work has expanded to include the entire productive process and it has become a means for organizing, managing and carrying out integral development projects for the communities. To fight the declining price crisis, farmers were compelled to find alternatives to conventional coffee production and marketing, so they decided on organic coffee production. The conversion to organic agriculture began in 1992, and by 1995 the first organic certificate was granted. The introduction of organic techniques has been carried out through the training of community promoters who create experimental organic lots in each community as a base for the learning process and research.

Farmers' extension covers the following themes i.e. soil conservation; production of organic fertilizers; coffee pruning; management of the diversity of animal and plant species; natural control of pests and diseases; organic production of crops for self-sufficiency in maize, beans and other food species; and internal control for supervision of the organic certification of the coffee and quality control of the product.The management of biodiversity within the coffee production systems and other cultivations constitutes an example of a rich local germplasm and of knowledge applied in the design of the stratification of the vegetation. This is knowledge transmitted from generation to generation resulting from a continuous process of adaptation. In 1995, a census was carried out on the species found in the organic coffee production systems of La Unión. The study demonstrated that besides coffee, there were more than thirty associated plant species of agronomic interest: fruit trees (loquat, mango, lime, guava, peach and orange), shade trees (eucalyptus, ash and pine), horticultural crops (tomato, chilli and beans), as well as medicinal plants and others used for the prevention of erosion. The benefits generated by the higher organic coffee prices were therefore accompanied by an improvement in the biodiversity of the coffee production system.

Thus, the polycultures systems established by these communities (like many others around the world) are characterized by highly diversified ecosystems and an improved agricultural biodiversity. The good markets associated with organic products has not only provided food but has also generated further community services.

1.14 Organic Success

1.14.1 Monsanto Ordered to Pay Millions in Landmark Case

In August 2018, Monsanto was ordered to pay $78 million in damages to DeWayne Johnson, who argued that exposure to the Roundup herbicide he sprayed while working as a school groundskeeper caused him to develop non-Hodgkin lymphoma.The jury's unanimous decision said Monsanto's products presented a "substantial danger" to people and the company failed to warn consumers of the risks. Lee Johnson is one of more than 4,000 people from across the country to file suit against Monsanto in state and federal courts based on allegations linking Roundup to cancer. After this ruling, we can assume that these other cases will move forward with renewed vigor.

1.14.2 The Organic Sector Grew to Almost $90 Billion

The 2018 edition of the study "The World of Organic Agriculture" shows that consumer demand for organic products is increasing, more farmers cultivate organically, more land is certified organic, and 178 countries report organic farming activities.The global market for organic food reached $89.7 billion. The United States is the leading market, worth 38.9 billion euros, followed by Germany, France, and China. In 2016, most of the major markets continued to show double-digit growth rates, and the French organic market grew by 22%. The highest per capita spending was in Switzerland (274 Euros), and Denmark had the highest organic market share (9.7% of the total food market). The countries with the largest organic share of agricultural land of their total farmland are Liechtenstein (37.7%), French Polynesia (31.3%), and Samoa (22.4%).

1.14.3 European Court of Justice Rules that New Genetic Engineering Techniques Must be Regulated as GMOs

In July 2018, the Court of Justice European Union (CJEU) ruled that new breeding methods such as CRISPR and other gene editing techniques are forms of genetic engineering and that products therefrom are indeed Genetically

Modified Organisms (GMOs) that should be regulated as such under existing EU legislation.

1.14.4 Denmark to Invest One Billion Kroner in Organic

In April 2018, the Danish government presented a new financial growth plan for organic agriculture worth 1.1 billion kroner (147 million euros). The stimulus package is largely designed to increase the number of farmers who choose to go organic. A total of 1.1 billion kroner in total will be spent in 2018 and 2019 in order to help farmers to convert to organic production.

1.14.5 India's Supreme Court Rules that Seeds, Plants and Animals not Patentable

In May 2018, India's Supreme Court refused to grant a stay on a ruling by Delhi's High Court that Monsanto cannot claim patents for its genetically modified cotton seeds Bollgard and Bollgard II in India.Monsanto first introduced its GMO-technology in India in 1995. Now, more than 90 percent of the country's cotton crop is genetically modified.

1.14.6 Sikkim wins Future Policy Award

The Indian state of Sikkim won the Future Policy Award, which this year celebrated the world's best policies on agroecology. In January 2016, Sikkimbecame India's first "100% organic" state. Today, all farming in Sikkim is carried out without the use of synthetic fertilizers and pesticides, providing access to safer food choices and making agriculture a more environment-friendly activity.

1.14.7 More than 69.8Mha of Organic Farmland – Australia has the largest Area

In 2017, 69.8Mha of organic agricultural land, including in-conversion areas, were recorded. The regions with the largest areas of organic agricultural landare Oceania (35.9Mha, which is half the world's organic agricultural land) and Europe (14.6Mha, 21%). Latin America has 8Mha (11%) followed by Asia (6.1Mha, 9%), North America (3.2Mha, 5%), and Africa (2.1Mha,3%). The countries with the most organic agricultural land are Australia (35.6Mha),Argentina (3.4Mha), and China (3Mha). Almost a quarter of the world's organic agricultural land (16.8Mha) and more than 87% (2.4Mha) of the producers were in developing countries and emerging markets.

1.14.8 Globally, 1.4% of the farmland is organic – Liechtenstein has the highest organic share with 37.9%

Currently, 1.4% of the world's agricultural land is organic. The highest organicshares of the total agricultural land, by region, are in Oceania (8.5%) and Europe(2.9%; European Union 7.2%). However, some countries reach far highershares: Liechtenstein (37.9%) and Samoa (37.6%) have the highest organicshares. In fourteen countries, 10% or more of the agricultural land is organic.

1.14.9 Record Growth in Organic Farmland - Increase of 11.7 Mha or 20%

Organic farmland increased by 11.7Mha or 20% in 2017, the largest increase ever recorded. The strong increase is mainly because 8.5 Million additionalhectares were reported from Australia. However, many other countries reported an important increase and thus contributed to the global growth, such as China(32% increase; over additional 0.7Mha), Argentina (12% increase; more than additional 0.4Mha), and the Russian Federation and India, both with an additional 0.3Mha. There was an increase in organic agricultural land in all regions. In Europe, the area grew by almost 1Mha (7.6% increase). In Asia, the area grew by almost 30% or an additional 1.2Mha; in Africa, the area grew by 14% or over 0.2Mha; inLatin America the area grew by 7% or 0.5Mha; and in North America by more than 3% or almost additional 0.1Mha. Apart from the organic agricultural land, there is organic land dedicated to other activities, most of which are areas for wild collection and beekeeping. Other areas include aquaculture, forests, and grazing areas on non-agricultural land. These areas of non-agricultural land constitute more than 42.4Mha.

1.14.10. Organic producers on the rise – 2.9 million producers in 2017

There were at least 2.9 million organic producers in 2017. Forty percent of the world's organic producers are in Asia, followed by Africa (28%) and Latin America(16%). The countries with the most producers are India (835'000), Uganda(210'352), and Mexico (210'000). There has been an increase in the number of producers of over 100'000, or nearly 5%, compared to 2016.

1.15 The Fastest Growing Green Startups

Green startups face additional challenges brought on by their inherent "triple bottom line," social responsibility, economic value, and environmental impact. Green startups are attracting venture capitalists and are gaining new support from socially responsible investors, green investors, and popular crowd funding platforms. From healthy fast food to organic lawn care and alternative energy to grassroots organizations, here's a list of 10 of 2018's fastest growing green startups. These 10 green startups have built their businesses on a platform for a greater, greener world.

1.15.1 Impossible Foods Inc

"We use plans to make the best meats and cheeses you'll ever eat," boasts Impossible Foods Inc., a startup offering sustainable vegetarian produce to conscious consumers across the United States. Impossible Foods was founded in 2011 out of Redwood City, California. By a complex molecular process, Impossible Foods selects specific proteins and nutrients from green foods to create imitation foods including cheeseburgers.

1.15.2 Choose Energy

The Texas-based startup, Choose Energy, was founded in 2008 as a marketplace for clean technology and services. Choose Energy, launched with the vision of simplifying shopping for electricity and natural gas rates, plan terms and renewable energy options for the average consumer and small business. Choose Energy makes it easy for people to save money on energy costs by switching to a 100% green product such as wind power

1.15.3 Holganix

Holganix, "Organic lawn care," sells environmentalfriendly lawn care products that contain natural microorganisms. Their products reduce the use of fertilizers as much as 90%.

1.15.4 Elevate Structure

Elevate Structure was launched in 2012 by a team of residential engineers in Hawaii with a dream of developing profitable real estate by building eco-friendly structures.

1.15.5 Solarkiosk

Solarkiosk, a Berlin-based social enterprise has provided solar-powered autonomous business hubs to off-the-grid communities since 2011. By 2014, Solarkiosk had established six subsidiaries in different countries across Africa and Asia. Rural communities now have access to sustainable energy, refrigeration, water purification, charging, communication, technology, information and business opportunities previously unheard of. This bottom of the pyramid approach enables and empowers local kiosk owners with green technology.

1.15.6 Freight Farms

Freight Farms, a Boston-based startup launched in 2010, develops shipping containers turned into self-contained farms. Freight Farms enables customers to grow fresh products using LEDs and hydroponics in any environment 365 days a year.

1.15.7 WISErg

WISErg describes itself as a "hybrid technology company combining bio-, clean-, and high-tech systems to create a revolutionary solution for managing urban-generated organics." The Harvester product is a machine that transforms food waste into high-quality fertilizer before it becomes waste. The food matter becomes a sort of high-nutrient liquid that can be converted to organic lawn care products. Launched commercially in 2014, the company achieved early success. Harvesters are installed in stores and facilities including Whole Foods Market.

1.15.8 Mow Green.US

MowGreen.US is a quiet, gas-free lawn care startup launched out of Connecticut by a local team of advocates for alternative energy and environmental awareness.

1.15.9 Arctic Sand

Arctic Sand launched out of the Massachusetts Institute of Technology (MIT) in 2011 with the goal of saving 80% of all energy that is lost in the form of heat worldwide. The power conversion technology startup offers products with proprietary technology developed and exclusively licensed from MIT.

1.15.10 Skeleton Technologies

Skeleton Technologies won the Ecosummit London Award for the best startup in 2015, as Europe's biggest developer and manufacturer of ultra-capacitor cells. Customers include the European Space Agency, which uses the product to solve power delivery and energy storage issues by recapturing the energy and providing peak and backup power.

References

ASSOCHAM (2018). The Indian Organic Market A New Paradigm in Agriculture.

Blackmore, S. (1994). Precision Farming: An Introduction. Outlook Agric. 23(4): 275–280.

FiBL and IFOAM (2019). Organics International- the World of Organic Agriculture, Statistics & Emerging Trends Edited by Helga Willer and Julia Lernoud.

Garibaldi, L.A.; Gemmill-Herren, B.; D'Annolfo, R.; Graeub, B.E.; Cunningham, S.A.; Breeze, T.D. (2017). Trends Ecol. Evol. (Amst.) 32(1): 68–80.

Gold, M.V. (2007). Sustainable Agriculture: Definitions and Terms. Related Terms. Available at. https://www.nal.usda.gov/afsic/sustainable-agriculture-definitionsand-terms-related-terms.

http://www.fao.org/agriculture/crops/core-themes/theme/pests/ipm/en/

https://en.wikipedia.org/wiki/History_of_organic_farming

Kovács-Hostyánszki, A.; Espíndola, A.; Vanbergen, A.J.; Settele, J.; Kremen, C.; Dicks, L.V.; (2017). Ecol. Lett. 20(5): 673–689.

Krishna, P. (2020). Northeast May Become Hub for Organic Farming: PM Modi at ICCs Annual Plenary Session. Business World.

Kumar, N. (2020). Govt plans to boost organic farming by doubling allocation, but experts say it's anti Swadeshi. Business Today, 3rd July 2020. https://www.businesstoday.in/sectors/agriculture/govt-plans-to-boost-organic-farming-by-doubling-allocation-but-experts-say-its-anti-swadeshi/story/408864.html

Lai, R. (1989). Adv. Agron. 42: 85–197.

Lançon, J.; Wery, J.; Rapidel, B.; Angokaye, M.; Gérardeaux, E.; Gaborel, C.; Ballo, D.; Fadegnon, B. (2007). Agron. Sustain. Dev. 27(2): 101–110.

Leakey, R.R.B. (2014). Annu. Rev. Phytopathol. 52: 113–133.

Linking Environment and Farming (LEAF) (2017). Integrated Farm Management. Available at https://leafuk.org/farming/integrated-farm-management.

Panwar, J.D.S.; Jain, A.K. (2015). Organic Farming- Scope and Use of Biofertilizers, New India Publishing Agency New Delhi

Plumecocq, G.; Debril, T.; Duru, M.; Magrini, M.B.; Sarthou, J.; Therond, O. (2018). Ecol. Soc. 23(1): 21.

Pretty, J.; Bharucha, Z.P. (2014). Ann. Bot. 114(8): 1571–1596.

Rana, S.S. (2016). Organic Farming Department of Agronomy, COA, CSK HPKV, Palampur, HP.

Rosea, David C.; William J. Sutherlandb, Andrew P.; Barnesc, Fiona Borthwickc, Charles Ffoulkesd; Clare Hallc,e, Jon M. Moorbyf, Phillipa Nicholas-Daviesf, Susan Twiningd,G.; Lynn V. Dicksh (2019). Land Use Policy 81: 834–842.

Yadav, A.K. (2011). Organic Agriculture – concepts, principles and practices. National Centre of Organic Farming, Department of Agriculture and Cooperation, Ministry of Agriculture, Govt. of India, Ghaziabad, Uttar Pradesh.

2

Biofertilizers

2.1 Introduction

Microbes are the oldest form of the life on earth. These organisms date more than 3 billion years to a time when they were covered with oceans. They are so tiny in size that million microbes can fit into the eye of a needle. Without microbes, we can't live. Thus, understanding microbes is vital to understanding past and future. Microbes play vary roles in the earth's environment recycling dead plant and animal matter through the soil, removing CO_2 from the atmosphere by photosynthesis in the oceans and nitrogen from the atmosphere to form nitrogenous fertilizers plants. Microbes play an important role in nutrient recycling, management, organic matter, decomposition and increasing the crop productivity and quality against diseases. Over the past century, many of the basic scientific principles of plant nutrition and soil fertility have been explained. In developed countries, inorganic chemical fertilizers have been widely accepted as a major source of improving and maintaining soil fertility. Plants need sufficient nutrients in proper balance for normal growth and development. Seventeen plant nutrients are essential for proper crop development. Each is equally important to the plant, yet each is required in vastly different amounts. Our understanding of nutritional needs of crop plants begin from the famous five year willow tree experiment conducted by J.B. van Helmont (1579–1644). He observed that willow tree had gained nearly 74.4kg but the loss of soil was only 56g, and concluded that the tree drew its nutrients from water not soil. Now we are aware that the soil is the repository for most of the plant nutrients, hence the concern for its continued health and sustainability (Brahmaprakash and Sahu, 2012). Soil environment needs to be made congenial for living of useful microbial population, responsible for continuous availability of nutrients from natural sources (Panwar and Jain, 2015).

2.2 Chronology of Major Developments Related to Biofertilizers

Although the beneficial effect of legumes in improving soil fertility known since ancient times, the field of biological nitrogen fixation opened up with the discoveries by Boussingault and Hellreigel in 1886. The commercial history of biofertilizers began with the launch of nitrogen by Nobbe and Hiltner, laboratory culture of rhizobia in 1895 followed by discovery of *Acetobacter* and then Blue green algae and a host of other microorganisms. In Indian, N.V. Joshi indicted first study on legume *Rhizobium* symbiosis and the first commercial production started as early as 1956. Biofertilizers offer a new technology for agriculture holding a promise to balance many of the shortcomings of the conventional chemical based technology. The main incentive for farmers to use biofertilizers seems to be that they hope to increase the yield or quality of their crops at a relatively low cost without a large investment of money. Nobbe Hiltner produced for the first time a laboratory culture of *Rhizobia* under the name Nitragin in 1885. Starting with *Rhizobia,* a vigorous search for other N fixing microorganisms began and soon it was found that there were non-symbiotic bacteria such as *Azotobacter*, which could fix atmospheric nitrogen. It was followed by a third group called blue green algae: Subsequently, microorganisms capable of solubilizing phosphate were discovered. In 1970, a new group of bacteria called *Azospirillum* identified. Mycorrhizae (AM) are the latest introduction in the list of biofertilizers which mobilize phosphorus. In India, systematic research on biofertilizers started with the first study of N.Y. Joshi in 1920. This was followed by extensive research by Gangulee, Sarkaria and Madhok on the physiology of the nodule bacteria and inoculation for increasing crop yields. Important milestones in production development and promotion of biofertilizers in India are given below.

Year	Some Milestones in Research, Production and Promotion of Biofertilizers in India
1834	Classical concept of Nitrogen Fixation by legumes "enunciated by French Agricultural Chemists J. B. Boussingault.
1886	Discovery of symbiotic nitrogen fixation by M. Hellriegel and H. Wilfarth, German Scientists production.
1888	Discovery that a bacterium (now named *Rhizobium*) is responsible for symbiotic nitrogen fixation in legumes by M. Beijerinck, a Dutch Scientist.
1895	Beginning of legume inoculants with the introduction of a biofertilizer named Nitragin by F. Nobbe and L. Hitner in the USA.
1905	Discovery of *Azotobacter* by Beijerinck.
1905	Start of biofertilizer production in Canada.
1912	Initiation of quality control legislation of biofertilizer in USA.
1914	Start of biofertilizer production in Australia and Sweeden.
1920	First study on legume –*Rhizobium* symbiosis in India by NV Joshi.
1925	Discovery of *Azospirillum* by M. Beijerinck.
1927	Initiation of mass culture of *Rhizobium* in fermenter vessel with aeration by Matchettle in USA.
1930	Issue of license to a biofertilizer manufactured by the US Government on the basis of testing culture.

Year	Some Milestones in Research, Production and Promotion of Biofertilizers in India
1932	Introduction of Yeast Extract Manittol Agar (YEMA) medium for *Rhizobium* by Fred.
1934	Earliest documented production of *Rhizobium* inoculants in India by M. R. Madhok.
1939	First report on the performance of *Azotobacter* in paddy/rice soil in India by B. N. Upal.
1939	Discovery of Blue Green Algae (BGA) as nitrogen fixer in paddy field by P. K. Dey.
1948	Recognition of peat as carrier in India.
1951	Use of charcoal as carrier in India.
1955	Use of Azolla as a biofertilizer in rice fields in North Vietnam.
1956	Use of Phosphate Solubilizing biofertilizers "Phosphobacterium" in the USSR.
1956	First commercial production of biofertilizers in India (New Delhi/Tamilnadu).
1957	Study on the solubilization of phosphate by micro-organisms in India by A. Sen and N.B. Pal.
1958	First attempt to standardize quality of legume inoculants in India by A. Sankaran.
1958	Recognition of *Azotobacter* culture as biofertilizers in the USSR with the trade name Azobacterin.
1960	First isolation of new non-symbiotic nitrogen fixing organism Derxia gummosa by P. K. Dey and Roma Bhattacharya of India.
1964	Spurt in demand of biofertilizer for soybean particularly in Madhya Pradesh.
1968	All India Co-ordinated Research Projects (AICRP) on Pulses improvement and soyabean set up by ICAR where *Rhizobium* study got priority.
1969	Use of Indian peat as carrier reported by V. Iswraan.
1970	Use of Charcoal, lignite and FYM as alternate carriers to peat reported by V. Iswraan.
1975	Coal as alternate carrier to peat reported by J.N. Dube.
1977	Indian Standard (ISI) for *Rhizobium* inoculants published.
1979	All Indian Co-ordinated Research Project on BNF initiated by ICAR, New Delhi.
1979	ISI standard for *Azotobacter* inoculants published.
1983	National Project on Development and Use of Biofertilizers by Ministry of Agriculture, GOI
1985	First national productivity award on biofertilizers given.
1988	Discovery of *Acetobacter* as Nitrogen Fixing bacteria in sugarcane by V.A. Cavalcante and J. Dobereiner in Brazil.
1988	National Facility for Blue Green Algae (BGA) Collection at IARI, New Delhi by Department of Biotechnology, GOI.
1988	National Research Centre for BGA set up at IARI, New Delhi.
1991	National Facility of *Rhizobium* Germplasm collection set up at IARI New Delhi.
1993	Mission Mode project on use and Development of biofertilizers technology set up by the GOI.
1993	Biofertilizer Newsletter published under the National Biofertilizers Project.
2001	National Bureau of Agriculturally Important Microorganisms (NBAIM) was established through a funded project sponsored by the Department of Agricultural Research and Education (DARE), Ministry of Agriculture, Government of India under ICAR, New Delhi.

Year	Some Milestones in Research, Production and Promotion of Biofertilizers in India
2006	Inclusion of four Biofertilizers i.e. *Rhizobium*, *Azospirillum*, *Azotobacter* and Phosphate Solubilizing Bacteria (PSB) in the Fertilizers Control Order (FCO) India.
2014	Technology for commercialization of Liquid Bio-Fertilizers
2016	Online PGS organic certification has been started under PGS-India programme through authorization of Regional Councils and capacity building of PGS local group by Hon'ble Union Minister of Agriculture Shri Radha Mohan Singh

Source: Panwar and Jain, 2015

2.3 Definition of Biofertilizers

Bio-fertilizers are natural fertilizers that are microbial inoculants of bacteria, algae, fungi alone or in combination. They are defined as a product containing carrier based (solid or liquid) living microorganisms that are agriculturally useful in terms of nitrogen fixation, phosphorous solubilization or nutrient mobilization. They augment the availability of essential elements like nitrogen, potash, phosphorous, sulphur by directly supplying them or transforming them into soluble form. In addition, they also help plants to uptake several micronutrients. Biofertilizers more commonly known as microbial inoculants are artificially multiplied cultures of certain soil organisms that can improve soil fertility and crop productivity. or

Bio-fertilizers being essential components of organic farming play vital role in maintaining long term soil fertility and sustainability by fixing atmospheric dinitrogen (N=N), mobilizing fixed macro and micro nutrients or convert insoluble P in the soil into soluble forms available to the plants, there by increases their efficiency and availability.

Biofertilizers have the ability to fix atmospheric nitrogen and mobilize phosphorus in soil from unavailable from the plant usable form. Biofertilizers include micro-organisms and their metabolites that are capable of enhancing soil fertility, crop growth and yield. These include both indigenous microbes and microbial inoculants that are microorganisms that replace fertilizers or increase a crop's fertilizer use efficiency. Thus, we can conclude that biofertilizers are ready to use live formulates of beneficial micro-organisms and their metabolites, which are capable of enhancing the soil fertility, crops growth and yield by mobilizing nutrients from no usable form to usable form through biological processes. An ideal fertile soil is characterized not only by optimum physical properties and chemical constituents conducive for plant growth, but also by a balance micro flora in the photosphere. With the introduction of green revolution technologies, there has been an increase in use of chemical fertilizers, pesticides, hybrid seeds, assured irrigation and agronomic practices that have disturbed the equilibrium of the ideal soil.

Due to this, there is an ongoing attempt on the part of Government of India to promote biofertilizers in Indian agriculture.

2.4 Classification of Biofertilizers

Biofertilizers can be defined as the preparations containing strains of microorganisms which can augment the microbiological process viz. nitrogen fixation, phosphate solubilization or mineralization, excretion of plant growth promoting substances or cellulose or lignin biodegradation in soil, compost or other environments. Biofertilizers can be divided into nitrogen fixing bacteria, phosphate solubilizing and mobilizing microorganisms and organic matter decomposers as (Table 2.1 and Fig. 2.1):

On the basis of Nitrogen Fixation: On the basis of N_2 fixation, they are divided in to two sub categories as

a) Symbiotic nitrogen fixers (*Rhizobium*, *Azolla* and *Frankia*)

b) Non-symbiotic nitrogen fixers (*Azotobacter*, *Azospirillum*, *Clostridium*, Blue Green Algae (BGA) as *Nostoc* , *Anaebaena* and *Calothrix*)

On the basis of **Phosphate uptake**: On the basis of Phosphate uptake, they are divided in to two sub categories as

a) Phosphate Solubilizing Biofertilizers (*Bacillus*, *Pseudomonas*, *Aspergillus*, and *Penicillium*)

b) Phosphate Mobilizing Biofertilizers (AM, Glomus, Acaluspora, *Gigaspora*, and *Scleroigha*)

On the basis of Organic Matter decomposition: On the basis of decomposition, they are divided in to two sub categories such as-

a) Cellulolytic biofertilizers

b) Lignolytic biofertilizers

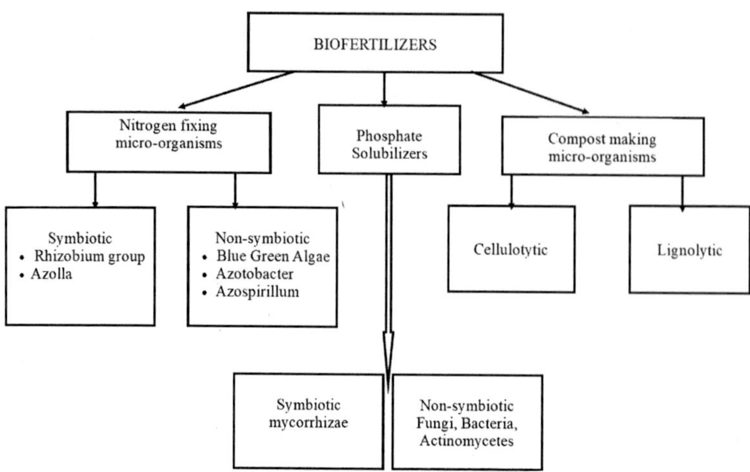

Fig. 2.1: Classification of Biofertilizers

Table 2.1: Different Clutches of Biofertilizers

S. No.	Clutches	Examples
N_2 Fixing Biofertilizers		
1	Free-living	*Azotobacter, Clostridium, Anabaena, Nostoc,*
2	Symbiotic	*Rhizobium, Frankia, Anabaena azollae*
3	Associative Symbiotic	*Azospirillium*
P Solubilizing Biofertilizers		
1	Bacteria	*Bacillus megaterium* var. *phosphaticum* *Bacillus circulans, Pseudomonas striata*
2	Fungi	*Penicillium* sps, *Aspergillus awamori*
P Mobilizing Biofertilizers		
1	Arbuscular mycorrhizae	*Glomus* sp.,*Gigaspora* sp.,*Acaulospora* sp., Scutellospora sps. and Sclerocystis sp.
2	Ectomycorrhiza	*Laccaria* sp., *Pisolithus* sp., *Boletus* sp., *Amanita* sp.
3	Orchid mycorrhiza	*Rhizoctonia solani*
Biofertilizers for Micro nutrients		
1	Silicate and Zinc Solubilizers	*Bacillus* sp.
Plant Growth Promoting Rhizobacteria		
1	Pseudomonas	*Pseudomonas fluorescens*

2.4.1 Nitrogen Fixing Biofertilizers

They are the preparations generally used as seed inoculants for the supply of nitrogen to the crops. Several heterotrophic bacteria have capacity to fix nitrogen from atmosphere and make it available for plant use.

2.4.1.1 Symbiotic Biofertilizers

Rhizobia

Rhizobia, Gram (-) soil bacteria belongs to family Rhizobiaceae, symbiotic in nature, fix nitrogen 50-100kg/ ha with legumes only. It is useful for pulse legumes like chickpea, red-gram, pea, lentil, blackgram, etc., oil-seed legumes like soybean and groundnut and forage legumes (berseem and lucerne). They are able to enter into symbiotic relationship with pulses and forms root nodules through the symbiosis by providing nitrogen to the host plants. The *Rhizobia*-legume association could fix up to 40- 200kg N/ha/season able to meet up to 80-90% of the crop nitrogen needs. They fix atmospheric nitrogen, and thus not only increase the production of the inoculated crops, but also leave a fair amount of nitrogen in the soil; which benefits the subsequent crop different legumes fix different amount of nitrogen depending on the crop, soil and environmental conditions. The efficiency of *legume Rhizobium* symbiosis depends upon the host crop and its compatibility with the *Rhizobium* bacteria, crop management and several soil factors like available soil nitrogen, soil moisture, soil pH and soil temperature etc. Successful nodulation of leguminous crops by *Rhizobium* largely depends on the availability of compatible strain for a particular legume as in mungbean (Jain *et al.,* 2007, 2008) and black gram (Rathi *et al,* 2009). It colonizes the roots of specific legumes to form tumor like growths called root nodules, which acts as factories of ammonia production. *Rhizobium* has ability to fix atmospheric nitrogen in symbiotic association with legumes and certain non- legumes like Parasponia. Following seven cross inoculation groups of *Rhizobia* have been recognized according to their host: (i) *Bradyrhizobium* spp. *Parasponia* (a non-legume, formally called Trema), (ii) *Bradyrhizobium japonicum* (Slow growing) Soybean (*Glycine max*), (iii) *Sinorhizobium fredii* (fast growing type) Alfalfa, (*Medicago sativa*), (iv) *Azorhizobium* (forms both root and stem nodules; the stem have adventitious roots) Sesbania (Aquatic), (v) *Rhizohium leguminosarum* by. *Phaseoli, Beans* (*Phaseolus*) (vi) *Rhizohium leguminosarum* by. *Viciae* Clovers (vii) *Photorhizobium* (Photosynthetically active).

Azolla—Anabaena symbiosis

Azolla is a small-leaf floating fern, which contains an endosymbiontic community living in the dorsal lobe cavity of the leaves. The presence in this cavity of a nitrogen fixing filamentous *cyanobacteria* in to the only fern-*cyanobacteria* association that presents agricultural interest by the nitrogen input that this plant could introduce in the fields. The nitrogen content of *Azolla* on dry weight basis is generally 5.0%. The fern can be used in rice crop by two ways as (i) green manure or (ii) dual crop. In case of green manuring, the fern is grown in the field or in separate shallow ponds and in corporate into soil prior to transplanting of rice. As a dual crop it is intercropped with rice and then mixed in the soil while crop is growing. It can fix about 30-40 kg N/ha/crop of *Azolla* that takes about 253 days. The limitation in use of *Azolla* as biofertilizer is scarcity of water and high temperature in north India. Higher temperature than 40°C is harmful for the fern maintenance of *Azolla* cultures in pond during winter is also a big problem. In India, for paddy fields *Azolla* is a promising biofertilizer and extensively used by farmers in several rice growing countries such as Philippines, China, Vietnam, Thailand and Sri Lanka. Apart from nitrogen fixation, Azolla is also known to suppress weed population in wet land rice and hence provides an additional economic advantage to rice cultivation. Recently, *Azolla* was also used as feed to enhance milk production in milching animals. *Azolla microphylla* (15t/ha) increases grain yield by 29.2% with neem cake reported by Sundaravarathan and Kannaiyan (2002).

Frankia

An actinomycetes is known to develop association with non-leguminous plants viz. Alnus, Casuarina, Ceanothus, Eleagnus, Myrica, Hippophae, and Purshia etc. and the causal organism is named as Frankia. It showed non-leguminous symbiosis. Nearly 120 species, mainly trees and shrubs are now known to harbour N_2 fixing bacteria in their root nodules which are not rhizobia. The non-leguminous nitrogen fixing systems are of modest agricultural importance, because their potential is not widely known. The extent of nitrogen gain by such Angiosperms varies enormously depend on soil type, climatic conditions and plant age. Nitrogen gains in the range of 12-200kg/ha/yr was reported for *Alnus* species, and 27-179kg/ha/yr for *Hippophae* species. Tests of individual location suggest the nitrogen fixation of 58kg/ha/yr for *Casuarina,* 60 kg/ha/yr for *Ceanothus* but only about 10kg/ha/yr for *myrica.* Plants in the group have little agricultural potential as grain crops or forage, but *Alder* could be available in agro-forestry. They may be used for upgrading poor soils. Its role as biofertilizer is insignificant.

2.4.1.2 Non-Symbiotic Nitrogen Fixers

***Acetobacter* inoculant**

Acetobacter diazotrophious (endophytic diazotrophs) isolated from leaves, stems and roots of sugarcane. In the eighties Dobereiner and coworkers in Brazil discovered this association and named *Acetobacter diazotrophicus*. These live in xylem vessels, intercellular space of root, shoot or leaf, ensuring proper supply of nutrients for nitrogen fixation. It was widely studied and used as a model system to assess the bacterial endophyte plant interaction. After its discovery, it was reported from a variety of crops like coffee (Jimnez-Salgado *et al.,* 1997), ragi (Loganathan *et al.,* 1999) and pineapple (Herandez *et al.,* 2000). and a latest report throw light on *Gluconacetobacter* sp as a natural colonizer of the wild rice (*Porteresia cocarctata Tateoka,* formerly *Oryza coarctata Roxbi*) and a salt tolerant pokali rice variety (Loganathan *et al.,* 2003). In India, its nitrogen fixing ability and plant growth promoting rhizobacteria (PGPR) production ability have been described, tested and found promising (Saravanan *et al.,* 2008). Besides, the bacterium secretes plant growth promotory substances such IAA and favours the germination and root development, which in turn help in the absorption of plant nutrients effectively from soil. *A. diazotrophicus* could enhance- rice growth and that growth promotion- might be related to the transfer of biologically fixed N although other factors such as auxin production could be involved reported by Sevilla and Kennedy *et al.,* (2000).

Gluconoacetobacter diazotrophicus colonizes the internal root tissue of sugarcane and fix nitrogen. Since it occupies vascular tissue, it has the obvious advantage of being first in line and thus solves the problem of competition by non diazotrophs. *G diazotrophicus* is also found in *Pennisetum purpureum, Ipomoea batatas* and *Arabica* (non-legume plants). The largest effect in this group was obtained with sugarcane, which can obtain up to 150kg N/ha from BNF (Dobereiner, 1997). In Tamilnadu, 24 strains of *G. diazotrophicus* were isolated from sugarcane (root, stem and leaves) and screened for nitrogenase activity. The strain isolated from sugarcane stem showed maximum acetylene reduction assay, 410.92n moles of C_2H_4/hr/mg cell protein followed by SoL3 from sugarcane leaf and hence the strain from sugarcane stem was recommended as an effective biofertilizer (Meenakshisundaram *et al.,* 2010).

Azospirillum

Azospirillum, a member of Spirillaceae, is associative N_2 fixing microorganism beneficial for non-leguminous plants. Associative diazotrophs are those in which is some in independence between the partners but they can grow satisfactory apart. They are found in association with cereals, grasses, vegetables, and oil seeds either freely in soil or indeed the root system. They not only fix nitrogen but also benefit plants by supplying growth hormones and vitamins. *Azospirillum* with farmyard manure (FYM), led to saving of 15-25kg equivalent of nitrogen also the above ground portion of plant through associative symbiosis. *Azospirillum* inoculation helps the plants in better vegetative growth saving nitrogenous fertilizers by 25-30% and fix nitrogen from 10 to 40kg/ha. *Azospirillum are* micro-arophilic in nature (Panwar and Sirohi, 1989). Following three species of *Azospirillum* have been identified as (i) *A. lipoferum,* (ii) *A. brasilense, and* (iii) *A. amazonenes. Azospirillum brasilense* is abundant in Indian soil. *Azospirillum* is often used with Mycorrhizal fungi for optimum growth and yield. One of the characteristics of *Azsopirillum* is ability to reduce and denitrify, therefore it is these characters. Inoculations with *Azospirillum* have registered increased in different vegetable crops.

It also secretes phytohormones in the plant root regions, which intern enhances the root growth. In developing countries like India, the use of *Azospirillum* as a biofertilizer would not only to the nitrogen supplementation to crops but also help in improving the fertility of soil in the long run. Its application is common practice in vegetables in Tamil Nadu. Increased yield in pearl millet obtained by inoculation of *Azospirillum* was due to production of IAA, gibbelellins, any cytokinine like substances by the bacterium and their subsequent effect on the plant (Kumar *et al.,* 2010). *Azospirillum* spp. is not considered to be a classic bio-control agent of soil-borne plant pathogens. However, there have been reports on moderate capabilities of *A. brasilense* in bio-control of crown gall-producing *Agrobacterium* (Bakanchikova *et al.,* 1993), bacterial leaf blight of mulberry (Sudhakar *et al.,* 2000) and bacterial leaf and/ or vascular tomato diseases (Bashan and Bashan, 2002b). In addition, *A. brasilense can* restrict the proliferation of other nonpathogenic rhizosphere bacteria (Holguin and Bashan, 1996). These antibacterial activities of *Azospirillum* could be related to its already known ability to produce bacteriocins (Oliveira and Drozdowicz 1987) and siderophores (Tapia-Hernández *et al.,* 1990). *A. brasilense* can synthesize phenylacetic acid (PAA), an auxin like molecule with antimicrobial activity (Somers *et al.,* 2005). Biofertilizers made from *Azospirillum* is suitable for C_4 crops such as sugarcane, maize, bajra, sorghum.

2.4.1.3 Blue Green Algae (BGA)

BGA are also known as *Cyanobacterium* as species of *Anabena, Aulasira, Gioeoirieha, Nostoc, Plectonema* etc. *Cyanobacteria* enter into a much wider range of association than *Rhizobia* due to their ability to fix nitrogen independently of the environment provided by the plant; they can be seen as more likely candidates than Rhizobia to form productive associations with cereals. *Cyanobacteria* were also found in the intercellular spaces of paranodules tissue in agreement with findings for *Azorhizobium* (Yu and Kennedy, 1995) and *Azospirillum* (Zemen *et al.,* 1992). Many of them are nitrogen fixing and synthesize their own energy source for nitrogen fixation and release it into surrounding in the form of amino acids, proteins and other growth promoting substances. Low land rice is an ideal ecosystem for their survival, growth and nitrogen fixation and therefore they can fix about 20-30-kg N/ha/season. Application of 1kg BGA culture in rice can increase the yield up to 10%-under optimum conditions. These belongs to eight different families, phototrophic in nature and produce auxin, indole acetic acid and gibberellic acid, fix 20-30kg N/ha in submerged rice fields as they are abundant in paddy, so also referred as **'Paddy Organisms'**. N is the key input required in large quantities for low land rice production. Soil N and BNF by associated organisms are major sources of N for low land rice. The 50-60% N requirement is met through the combination of mineralization of soil organic N and BNF by free living and rice plant associated bacteria (Roger and Ladha, 1992). Most N fixing BGA are filamentous, consisting of chain of vegetative cells including specialized cells called heterocyst which function as micro nodule for synthesis and N fixing machinery.

2.4.2. On the basis of Phosphate Solubility/Mobility

Phosphorus differs fundamentally from nitrogen in that no natural channels exist for return of large amount of losses occurring annually. Hence, the supply is inevitable shrinking and deposits are limited. Farmers are constrained in using phosphorus owing to in high cost of phosphatic fertilizers. In such situations, the release of insoluble phosphorus in soil and fixed phosphorus in clay minerals by microorganisms assumes great significance. They may be broadly classified in two groups namely phosphate solubilizers and phosphate absorbers.

2.4.2.1 Phosphate Mobilizing Biofertilizers

P mobilizers facilitate the mobilization of soluble phosphorus from distant places in soil, where plant roots cannot reach and thus increase availability of P to plants. Mycorrhizae are prominent P mobilizers. Mycorrhizae are a symbiotic association between plant roots and a few fungus. The fungal partner is benefited by obtaining its carbon requirements from host's photosynthates and the plant in turn gains the much needed nutrients especially phosphorus, calcium, copper and zinc which would otherwise be inaccessible to the host. This uptake of nutrients is facilitated with the help of fine absorbing hyphae of the fungus. These fungi are associated with majority of agricultural crops. There are seven genera of these fungi that produce Arbuscular mycorrhizal symbiosis with plants. They are *Glomus, Gigaspora, Scutellospora, Acaulospora Entrophospora, Archaeospora* and *Paraglomus.* They account for 5–50% of the biomass of soil microbes (Olsson *et al.,* 1999). Hyphal biomass of AM fungi may amount to 54–900 kg/ha (Zhu and Miller, 2003), and some products formed by them may account for another 3000 kg (Lovelock *et al.,* 2004). Pools of organic carbon such as glomalin produced by AM fungi may even exceed soil microbial biomass by a factor of 10–20 (Rillig *et al.,* 2001). Approximately 10–100m mycorrhizal mycelium can be found per cm root (Mc Gonigle and Millerv, 1999). The mechanism that is generally accepted is a wider physical exploration of the soil by mycorrhizal fungi (hyphae) rather than by roots. A speculative mechanism to explain P uptake by mycorrhizal fungi involves the production of glomalin (Lovelock *et al.,* 2004). The micro-organisms always participate in cycling of phosphorus in the environment hence there is recycling and not exhaustion.

AM fungi play an important role in water economy of plants. Their association improves the hydraulic conductivity of the root at lower soil water potentials, and this improvement is one of the factors contributing towards better uptake of water by plants (Mahdi *et al.,* 2010). A few proposed mechanisms by which AM fungi also help in activation of plant defense systems include changes in exudation patterns and concomitant changes in mycorrhizosphere populations, increased lignifications of cell walls, and competition for space for colonization and infection sites (Kasiamdar *et al.,* 2001).

2.4.2.2 Phosphate Solubilizers

Phosphorus is second only to nitrogen in mineral nutrients most commonly limiting growth of crops, and an essential element for plant development and growth making up about 0.2% of plant dry weight. Plants acquire P from soil solution as phosphate anions. However, these are extremely reactive and may be immobilized through precipitation with cations such as Ca^{2+}, Mg^{2+}, Fe^{3+} and Al^{3+} depending on the particular properties of a soil. In these forms, P is highly insoluble and a large portion of soluble inorganic phosphate applied to the soil as chemical fertilizer is immobilized rapidly and becomes unavailable to plants (Goldstein, 1986). Hence, the amount available to plants is usually a small pro- portion of this total application. Several bacteria, particularly as *Bacillus polymyxa, Bacillus subtilis* and *Pseudomonas striata,* and fungi *as Penicillium digitatum* and *Aspergillus awamori* have the ability to convert the insoluble inorganic phosphorus into the soluble form, which can be utilized by crop plants. The inoculants of these microorganisms are called PSB (Phosphate solubilizing bacteria) or PSM (Phosphate solubilizing microorganisms) inoculants. They also produce siderophores (iron chelating substances, e.g. Pseudobactin), which chelate with iron and make it unavailable to harmful fungi as *Eriwina* in rhizosphere, and produce plant growth hormones like indole acetic acid and gibberellic acid etc. These cultures can be used in all crops of legumes, cereals and vegetables.

2.4.3 Phosphorous Biofertilizers

Phosphorous is very often present in the soil in unavailable form. Several soil bacteria particularly those belonging to the genera Bacillus and Pseudomonas possess the ability to bring insoluble phosphates in the soluble forms by secreting organic acids. These acids lower the pH and bring about the dissolution of bound forms of phosphorous. These bacteria are commonly known as phosphobacteria. They can be applied either through seed or soil application. Phosphorous is one of the important elements for plant growth. It is the second major chemical fertilizer, being applied for crop production. For optimum crop production such as wheat, soil level should be maintained above 30ppm available P in the top 15cm. It is anticipated that only 10-15% recovery is obtained for applied P fertilizers by plants. The remainder of the fertilizers-P is accumulating in the soil. This is available in two major forms: inorganic soil 'P' which comprises adsorbed phosphate and precipitated Fe, Al, $MnFePO_4$, $AlPO_3$, $Ca_3(PO_4)_2$ acidic soil and Ca salts e.g. $Ca_3(PO_4)_2$ alkaline soils. Another general form of unavailable soil phosphorus is organic soil P, which comprises in phosphates, phospholipids, nucleic acids etc. It occurs in inorganic as well as organic form, inorganic forms are compounds of calcium, iron, and aluminium in rock phosphates and organic forms are phytin, phospholipids, nucleic acids, mositol phosphates which are added to the soil by decaying vegetation. Rock phosphate is the raw material for the production of phosphate fertilizers. Phosphate biofertilizers were first prepared by USSR using *Bacillus magaterium* var. *Phosphaticum* as phosphate solubilizing bacteria and product was termed as phosphobacter.

2.4.3.1 Mycorrhizae

The term Mycorrhizae denotes "fungus roots". It is a symbiotic association between host plants and certain group of fungi at the root system, in which the fungal partner is benefited by obtaining its carbon requirements from the photosynthates of the host and the host in turn is benefited by obtaining the much needed nutrients especially phosphorus, calcium, copper, zinc etc., which are otherwise inaccessible to it, with the help of the fine absorbing hyphae of the fungus. These fungi are associated with majority of agricultural crops, except with those crops/plants belonging to families of Chenopodiaceae, Amaranthaceae, Caryophyllaceae, Polygonaceae, Brassicaceae, Commelinaceae, Juncaceae and Cyperaceae. They are ubiquitous in geographic distribution occurring with plants growing in artic, temperate and tropical regions alike. AM occurs over a broad ecological range from aquatic to desert environments (Mosse *et al.*, 1981). Of 150 species of fungi that have been described in order *Glomales* of class *Zygomycetes,* only small proportions are presumed to be *mycorrhizal.* There are 6 genera of fungi that contain species, which are known to produce *Arbuscular mycorrhizal fungi* (AMF) with plants (Panwar and Thakur, 1995). Two of these genera, *Glomus* and *Sclerocytis* produce chlamydospores only. Four genera form spores that are similar to *Gigaspora, Scutellospora, Acaulospora* and *Entrophospora.* The oldest and most prevalent of these associations are AM symbioses that first evolved 400 million years ago, coinciding with the appearance of the first land plants. In comparison, crop domestication is a relatively recent event, beginning 10,000 years ago (Sawers *et al.*, 2007). *Mycorrhizal* fungi can increase the yield of a plant by 30-40%. Plants that suffer from nutrient scarcity, especially, phosphorus develops m*ycorrhizae.* Association of mycorrhizae is wide spread in bryophytes a large number of pteridophytes, most or all species of Gymnosperms and some 90% or more of angiosperms. In recent years use of artificially produced inoculums of mycorrhizal fungi has increased due to its significance and multifarious role as

- Benefits more than 99% of earth's plants.
- Occupy 100 times more soil volume than a non-mycorrhizal plants entire root system.
- Increase absorptive surfaces or root systems up to 700%.
- Extend through the soil up to 30ft away from a plant host. Depress many root diseases caused by pathogenic fungi and nematodes. Develop tolerance to climatic and edaphically stresses including high soil temperature and heavy metal toxicity.

Ectomycorrhizae

It is also called as *Ectotrophic mycorrhizae* and found in approximately 10% of the world flora. It colonizes outside of plants cells and a root such as the fungus completely encloses each feeder rootlet in a sheath of mantle of hyphae, which penetrate only between the cells of root hairs. But due to the presence of layer of hyphae it almost looks like a host tissue and this layer is known as pseudoparendchymatous sheath. From this sheath hyphae enter the cortex and remain only in the outer cortical cells to form a network called as Hartig net. All the nutrients are absorbed by the fungal mantle and transported to the root through the hartig net. The fungi involved in *ectomycorrhizae* come under *Basidiomycetes* and they grow best at pH 5-6. The excess of inorganic fertilizers suppress *ectomycorrhizal* development and these results in the stunted growth and chlorosis. When the defense reaction of the higher symbiont diminishes, as it is likely to happen in senescent or diseased trees, the lower symbiont may become endotropic. Such instances have been designated as *ectoendomychorrhizae*.

Endomycorrhizae

Endomycorrhizae, the fungus does not form an external sheath but colonizes inside the plant root cells and establishes direct connections between the cells of the roots and surrounding soils. The fungi involved in endotropic association belong to either to the *Basidiomycetes* (possessing aseptate hyphae) or to the *Basidiomycetes* or fungi imperfecti (possessing septate hyphae). Root colonization involves the formation of intracellular and highly branched hyphae scattered throughout the root system. The establishment of an intracellular symbiosis between the fungus and the host roots requires the penetration of the cell by fungal cell wall hydrolyzing enzyme. Fungal hyphae enters the cells of the host plant and thus penetrates the cell wall at the site of contact with the aid of various hydrolytic enzymes like xylanases, pectinases and celluloses, the fungi can break down lignin and cellulose and thus contribute to the decaying of organic matter. In this respect, they differ from *ectomycorrhizal* fungi on the host for their carbon nutrition. For *ectomycorrhizal* fungi the basidiospores, pure mycelial culture, chopped sporocarp, sclerotia and fragmented *mycorrhizal* roots or soil from *mycorrhizosphere* region can be used as inoculum. The inoculum is mixed with nursery soils and seeds are sown. Now a day's commercially available mycelial inoculum are also available in the market.

AM *(Arbuscular Mycorrhizae)*

AM fungi are the ubiquitous soil microbes found associated with most of the angiosperms, pteridophytes, and bryophytes but absent in plants which from only *ectomycorrhizae* (Pinaceae and Betulaceae) or the two other specific types of *endomycorrhiza* of Ericales and Orchidales. AM fungi form symbiotic association with a number of economically important crop plants. These fungi can improve the plant growth through increased uptake of phosphorous, sulphur, calcium and zinc. Further these fungi are known to enhance resistance to diseases and help the host plant to absorb more water under moisture stress conditions. AM develop special structures called as arbuscles and vesicles that help in the transfer of the nutrients especially phosphates from the soil into the root system. VAM has been reported from 1000 genera of plants representing from 200 families. Fungus associated with VAM is generally non-septate Zygomycetes belonging to the genera Clomus, Gigaspora, Acaulospora, Seerocystis and Endogone. The fungi being obligate biotrophs don't grow on synthetic media and hence are classified according to the morphological characteristics of the spore formed in the soil. AM increases the plant growth mainly by increasing the soil volume from which plant absorbs relatively less mobile nutrients such as Zn, P, and Cu etc., which are very important for optimum plant growth (Jain and Tomer, 2018). However, response to AM fungi may vary depending upon crop species, soil P levels and the presence of other microorganism in the root regions. The potential of Arbuscular mycorrhizal fungi as biofertilizers and bioprotectors to enhance the micro propagated plantlets. Besides AM can be used as a disease control agent as the fungus is said to be releasing such chemical compounds which have an inhibitory effect on the pathogens. Studies have shown that AM formed by *Gigaspora* exhibits the inhibitory effect on the development of pigeon pea blight caused by *Phytophthora drechsleri.*

2.4.4 Zinc Solubilizers

The nitrogen fixers (*Rhizobium, Azospirillum, Azotobacter),* BGA and Phosphate solubilizing bacteria (*Bacillus magaterium* and *Pseudomonas striata*) and phosphate mobilizing mycorrhizae have been widely accepted as bio-fertilizers. However, these supply only major nutrients but a host of microorganism that can transform micronutrients are there in soil that can be used as bio-fertilizers to supply micronutrients like Zn, Fe, Cu etc. Zinc being utmost important is found in the earth's crust to the tune of 0.008% but more than 50% of Indian soils exhibit deficiency of zinc with content must below the critical level of 1.5ppm of available zinc (Katyaland Rattan, 1993). The plant constraints in absorbing zinc from the soil are overcome by external application of soluble zinc sulphate ($ZnSO_4$). But the fate of applied zinc in the submerged soil conditions is pathetic and only 1-4% of total available zinc is utilized by the crop and 75% of applied zinc is transformed into different mineral fractions (Zn-fixation) which are not available for plant absorption (crystalline iron oxide bound and residual zinc). There appears to be two main mechanisms of zinc- fixation, one operates in acidic soils and is closely related with cat ion exchange and other operates in alkaline conditions where fixation takes by means of chemisorptions, (chemisorptions of zinc on calcium carbonate formed a solid-solution of $ZnCaCO_3$, and by complexation by organic ligands (Alloway, 2008). The zinc can be solubilized by micro-organisms i.e. *Bacillus subtilis, Thiobacillus thioxidans* and *Saccharomyces* sp. These microorganisms can be used as bio-fertilizers for solubilization of fixed micronutrients like zinc. The results have shown that a *Bacillus sp.* (Zn solubilizing bacteria) can be used as bio-fertilizer for zinc or in soils where native zinc is higher or in conjunction with insoluble cheaper zinc compounds like zinc oxide (ZnO), zinc carbonate ($ZnCO_3$) and zinc sulphide (ZnS) instead of costly zinc sulphate (Mahdi *et al.*, 2010).

2.4.5 Organic Biofertilizers:

Soil organic matter plays important role in the maintenance and improvement of soil properties. And derived to a large extent from residues and remains of the plants together with the small quantities of animal remains, excreta, and microbial tissues. Soil organic matter is composed of three major components i.e. plants residues, animal remain and dead remains of microorganisms. Various organic compounds are made up of complex carbohydrates, (Cellulose, hemicellulose, starch) simple sugars, lignins, pectins, gums, mucilages, proteins, fats, oils, waxes, resins, alcohols, organic acids, phenols etc. and other products. When plant and animal residues are added to the soil, the various constituents of the soil organic matter are decomposed simultaneously by the activity of microorganisms and carbon is released as CO_2, and nitrogen (as $NH_4 \rightarrow NO_3$) for the use by plants. Other nutrients are also converted into plant usable forms. This process of release of nutrients from organic matter is called mineralization. The insoluble plant residues constitute the part of humus and soil organic matter complex. The final product of aerobic decomposition is CO_2 and that of anaerobic decomposition are hydrogen, ethyl alcohol (CH_4), various organic acids and carbon dioxide (CO_2). Soil organisms use organic matter as a source of energy and food. The process of decomposition is initially fast, but slows down considerably as the supply of readily decomposable organic matter gets exhausted. Sugars, water-soluble nitrogenous compounds, amino acids, lipids, starches and some of the hemicellulases are decomposed first at rapid rate, while insoluble compounds such as cellulose, hemicellulose, lignin, proteins etc. which forms the major portion of organic matter are decomposed later slowly. Thus, the organic matter added to the soil is converted by oxidative decomposition to simpler substances which are made available in stages for plant growth and the residue is transformed into humus. The microbiology of decomposition/degradation of some of the major constituents (cellulose, hemicellulose, lignin, proteins etc.) of soil organic matter/plant residues are discussed here.

2.4.5.1 Cellulose decomposition

Cellulose is the most abundant carbohydrate present in plant residues/organic matter in nature. When cellulose is associated with pentosans (xylans and mannans) it undergoes rapid decomposition, but when associated with lignin, the rate of decomposition is very slow. The decomposition of cellulose occurs in two stages as

a) In the first stage the long chain of cellulase is broken down into cellobiase and then into glucose by the process of hydrolysis in the presence of enzymes cellulase and cellobiase, and

b) In second stage glucose is oxidized and converted into CO_2 and water.

1. $\text{Cellulose} \xrightarrow[\text{hydrolysis}]{\text{Cellulase}} \text{Cellobiose} \xrightarrow[\text{hydrolysis}]{\text{Cellulase}} \text{Glucose}$

2. $\text{Glucose} \xrightarrow{\text{Oxidation}} \text{Organic Acids} \xrightarrow{\text{Oxidation}} CO_2 + H_2O$

The intermediate products formed/released during enzymatic hydrolysis of cellulose (cellobiose and glucose) are utilized by the cellulose-decomposing organisms or by other organisms as source of energy for biosynthetic processes. The cellulolytic microorganisms responsible for degradation of cellulose through the excretion of enzymes (cellulase and cellobiase) are fungi, bacteria and actinomycetes.

2.4.5.2 Hemicelluloses decomposition

Hemicelluloses are water-soluble polysaccharides and consist of hexoses, pentoses, and uronic acids and are the major plant constituents second only in quantity of cellulose, and sources of energy and nutrients for soil microflora. The hydrolysis is brought about by number of hemicellulolytic enzymes known as "hemicellulases" excreted by the microorganisms. On hydrolysis hemicelluloses are converted into soluble monosaccharide/sugars (xylose, arabinose, galactose and mannose) which are further convened to organic acids, alcohols, CO_2 and H_2O and uronic acids are broken down to pentoses and CO_2. Various microorganisms including fungi, bacteria and actinomycetes both aerobic and anaerobic are involved in the decomposition of hemicelluloses.

2.4.5.3 Lignin decomposition

Lignin is the third most abundant constituent of plant tissues, and accounts about 10-30% of the dry matter of mature plant materials. Lignin content of young plants is low and gradually increases as the plant grows old. It is one of the most resistant organic substances for the microorganisms to degrade however certain Basidiomycetous fungi are known to degrade lignin at slow rates. Complete oxidation of lignin result in the formation of aromatic compounds such as syring aldehydes, vanillin and ferulic acid. The final cleavages of these aromatic compounds yield organic acids, carbon dioxide, methane and water. Microorganisms involved in the decomposition of organic matter are listed in the table 2.2.

Table 2.2: List of Micro-organisms, involved in decomposition of organic matter

Constituents	Microorganisms		
	Bacteria	Fungi	Actinomycetes
Cellulose	*Achromobacter, Bacillus, Cellulomonas, Cellvibrio, Clostridium, Cytophaga, Vibrio Pseudomonas, Sporocytophaga etc.*	*Aspergillus, Chaetomium, Fusarium, Pencillium Rhizoctonia, Rhizopus, Trichoderma, Verticillttm.*	Micromonospora, Nocardia Streptomyces, Thermonospora
Hemicellulose	*Bacillus, Achromobacter, Cytophaga Pseudomonas, Erwinia, Vibrio, Lactobacillus*	*Aspergillus, Fusarium, Chaetomium, Penicillium, Trichoderma, Humicola*	Streptomyces, Actinomycetes
Lignin	*Flavobacterium, Pseudomonas, Micrococcus, Arthorbacter, Xanthomonas*	*Humicola, Fusarium Fames, Pencillium, Aspergillus, Ganoderma*	Streptomyces, Nocardia

2.4.6. Mixed biofertilizers

Mixed biofertilizers containing a consortium of nitrogen fixers, phosphate solubilizers and plant growth promoting *Rhizobacteria (*PGPR) were found to promote the growth of cereals, legumes and oil seeds better and save nitrogen and phosphate fertilizers in crops. Response was better and there was 50% nutrient substitution in crops like rice and legumes depending upon soil type and yield level. The response of biofertilizers was better when used along with 75% chemical fertilizers.

2.4.7. Other biofertilizers

2.4.7.1 Potash biofertilizers

Some microbes can be used as potash biofertilizer as *Bacillus mucilaginous*, *Frateuria aurantia*, *Trichoderma harzianum* etc. Release of K through solubilization of muscovite, orthoclase and microline by *Bacillus mucilagenosus* has been reported (Sugumaran and Janathanam, 2007).

***Bacillus mucilagenosus*:** Slime forming micro-organisms have been found to be useful for weathering of K bearing minerals. *Bacillus mucilagensous*, belonging to Bacilliaceae family is gram +ve, rod shaped slime producing bacteria which produce nuclease, endonucleases, cellobiase, protease, ribonuclease and phosphomonoestoarase Phylogenic analysis indicates that *B. mucilagenosus* and *B. edaphicus* are member of the genus *Paenibacillus*.

Frateuria aurantia: *Frateuria aurantia* belongs to the Pseudomonaceae family. This bacteria is gram-ve, rod shaped, obligatory anaerobe. It has been found to be capable of mobilizing potash, and may enhance crop productivity in K-deficient soils (Ramarethinam and Chandra, 2006). *F. aurantia* is also known to solubilize calcium. Even though K-solubilizing/mobilizing biofertilizers are being produced and marked, their mode of action is still not fully clear and much more research is needed in this area.

2.4.7.2 EM-based biofertilizers

The concept of effective micro-organisms (EM) was developed by Prof. Teruo Higa in Okinawa, Japan. EM is a multi-culture of major micro-organisms and a consortium of photosynthetic bacteria, lactic acid bacteria, yeast, actinomycetes and fermenting fungi with those found in nature and capable of co-existing with other microorganisms. EM microbial solution can convert all waste into very good fertilizers within a short time. EM is an alternative technology which is useful in recycling and management of organic wastes.

2.4.7.3 Nano-based biofertilizers

Nano science is now contributing to our knowledge in different area including agriculture. Efforts are being made to produce 'Nano Biofertilizers' based on Nano biotechnology. It has been observed that some nanoparticles (Particles size of 10^{-9}m) if mixed with rock phosphate can enhance the solubilizing capacity of P solubilizing microbes. Efforts are also on going to utilize nanoparticles to achieve higher mobilization of zinc mediated by rhizobacteria for crop production. Nano-biofertilizers are being produced and marked in Malaysia (Bhattacharya and Tandon, 2012).

2.5 Microbes as PGPR Usefulness in Biotic and Abiotic Stress

PGPR strains *Bacillus, Azospirillum Azotobacter* and *Pseudomonas* spp. etc. are found to improve seed germination, root/shoot length, nodulation, biomass, yield and nutrient mobilization in various crops (Table. 2.3) The beneficial bacteria also protect the crops from the attack of soil borne pathogen among which *Fusarium* spp., *Rhizoctonia solani, Rhizoctonia bataticola* and *Scierotium rolfsii* are highly destructive. Pyrrolnitrin, Phenazine-1-carboxylic acid and 2,4 produced by *Pseudomonas* spp. were the major antibiotics involved.

2.5.1 PGPR: Usefulness under biotic stress

Co-inoculation of PGPR along with *Rhizobium* increased the root nodulation. The beneficial effects of co-inoculation are strain dependent. The basic mechanisms involved in this synergistic activity were not completely known and remains a challenge. One possibility is that PGPR, by altering the host's secondary metabolism and/or creating antibiosis in the rhizosphere, out compete the pathogens and eliminate competition of *Rhizobium* with deleterious microorganisms for colonization of the plant parts. Alteration of the plant flavonoid metabolism was proposed as another mechanism of synergistic activity of PGPR and *Rhizobia*. Inoculation of *Azospirillium* sps. produces large quantities of auxins and stimulates the formation of epidermal cells that become root hair cells or additional infection sites for rhizobial colonization. Unlike chemical fertilizers, PGPR exert a beneficial effect on rhizobial nodulation of legumes and should be exploited for the economic benefit of subsistence farming systems.

Table 2.3: Plant growth promoting rhizobacteria (PGPR) in enhancing the growth of crop plants.

Crop	*Bacterial isolate*	Effects on growth and yield
Wheat	*Bacillus subtilis*	Seed germination and grain yield
Pigeonpea	*Bacillus substilis* AF1	Increase in shoot & root length and biomass
Mungbean, Pea	*Azotobacter*	Increase in yield
Chick pea	*Azospirillum brasilense* *Azotobacter, Pseudomonas* sp.	Enhanced seed germination, root and shoot dry weight, yield and nodulation Number of shoots/plant, yield, seed protein
Soybean	*Bacillus cereus UW 85*	Increase in root growth, nodulation and yield
Faba bean	*Azospirillum brasilense*	Increase in growth of root and shoot and nodulation
Groundnut	*Bacillus subtilis* *Bacillus amyloliquefaciens* *Pseudomonas aeruginosa,* *Pseudomonas fluorescens*	Increase in root and shoot length, biomass and yield.
Sunflower	*Azospirillum, Azotobacter* *Pseudomonas*	Increase in yield
Black pepper, ginger, turmeric, spice crop	*Pseudomonas fluorescens* *Phosphobacteria*	Nutrient mobilization

PGPR's and their effectiveness has been established on foot rot and slow decline of black pepper and rhizome rot of ginger and dump rot of cardamom. Strains of *Pseudomonas fluorescens* were found to increase the growth and vigor of these crops apart from suppressing the soil borne diseases (Sarma *et al.,* 2000).

2.5.2 PGPR: Usefulness under abiotic stress

Microbial colonization of plant root is affected by biotic and abiotic factors as root exudates, competition, inorganic nutrients, soil pH and temperature etc. Among the agriculturally important abiotic stress, drought, heat, soil salinity and alkanity as the most detrimental and cause loss of crop productivity all over the world. Tripathi *et al.,* (1998) have studied in detail salinity stress responses in the plant growth promoting rhizobacteria *Azospirillum* sps. They have shown that in order to adapt to the fluctuations in soil salinity/ osmolarity, the bacteria accumulate compatible solutes such as glutamate, proline, glycine betaine, trehalose etc. Proline seems to play a major role in osmoadoptation. With increase in osmotic stress the dominant osmolyte in *Azospirillum brasilense* shifts from glutamate to proline. Accumulation of proline in *Azospirillum brasilense* takes place during uptake and synthesis.

Rangarajan *et al., (*2001) analyzed populations of *Pseudomonas* for their biochemical characters and genetic diversity using molecular tools including RAPD and PCR-RFLP. It was observed that increasing salinity caused a predominant selection of salt tolerant species, in particular *Pseudomonas pseudoalcaligenes* and *Pseudomonas alcaligenes,* irrespective of the host rhizosphere. Rangarajan *et al.,* (2003) have isolated indigenous Pseudomonas strain from rhizosphere of rice cultivated in the coastal agroecosystem and screened for *in vitro* antibiosis against *Xanthomonas oryzae* and *Rhizoctonia solani.* PSB that could tolerate high salinity have also been isolated (Kumar, 2003). P-starvation inducible quinoprotein glucose dehydrogenase (Gcd) has been shown to be responsible for the high gluconic acid secretion resulting efficient P solubilization.

Table 2.4: Summary of the Micro-organisms and their functions used as Biofertilizers

Micro-organisms	Function/Contribution	Limitation	Use for crops
Rhizobium	a) Estimated fixation of 30-200kg N/ha. b) 10-35% increase in yield of field crops c) Fixed N also benefits next crop	a) Fixes N only with legumes b) Not very effective in traditional area c) Needs optimum supply of P and Mo	All pulses, beans Legume oilseeds (soybean, groundnut) Forage legumes like clover, lucerne, others
Azotobacter and *Azospirillum*	a) Fixation of 20-25kg N/ha b) 10-15% increase in yield c) Production of growth promoting substances such as IAA, GA, vitamins etc.	a) Demands high organic matter b) Azospirillum is effective in C4 plants	Cereals, millets Cotton, Sugarcane Vegetables/others
Acetobacter (Glucono acetobacter)	a) Can fix 100-150kg N/ha. b) 10-20% increase yield c) Can solubilize P also.		
Cyanobacteria (BGA)	a) Fixation of 20-30kg N/ha b) 10-15% increase in yield c) Production of growth promoting substances.	Demands and high organic matter and specific C substrate.	Only for sugarcane

Micro-organisms	Function/Contribution	Limitation	Use for crops
Azolla-Anabaena	Can fix 30-100 kg N/ha 10-25% increase in yield	a) Poor survival at high temperature b) High demand for P	Only for flooded rice or as a green manure
	a) Can solubilize 20-30% of insoluble phosphate b) 10-20% yield increase	Demand high organic matter	For all soils and crops
P mobilizing BF (Mycorrhizae)	a) Can mobilize P for plant b) 10-20% increase in yield	Demand organic matter	Potentially for all upland field and tree crops.
K biofertilizer Frateuria aurantia B. muvilagenosus	Can release 10-15 % K from K bearing minerals	Details under research	Potentially for all soils and crops

Source: Panwar and Jain (2015)

References

Alloway, B.J. (2008). In: Zinc in soils and crop nutrition. Second edition, IZA and IFA publishers, Brussels, Belgium and Paris, France. pp. 21-22.

Bakanchikova, T.I.; Lobanok, E.V.; Pavlova-Ivanova, L.K.; Redkina, T.V.; Nagapetyan, Z.A. and Majsuryan, A.N. (1993). Mikrobiologiya. 62: 515–523.

Bashan, Y.; de-Bashan, L.E. (2002b). Applied and Environmental Microbiology 68: 2637–2643.

Bhattacharya, P.; Tandon, H.L.S. (2012). In: Biofertilizer Handbook, FDCO, New Delhi. pp.29.

Brahmaprakash. G.P.; Sahu, P.K. (2012). Journal of the Indian Institute of Science 92 (1): 37-62.

Dobereiner, J. (1997). Soil Biology and Biochemistry. 29:771–774.

Goldstein, A.H. (1986). American Journal of Alternate Agriculture 1:57–65.

Herandez, A.T.; Bustilos-cristales, M.R.; Jimnez-salgado, T.; Caballero-Mellado, J.; Fuentez-Ramirez, L.E. (2000). Microbial Ecology 39: 49–55.

Holguin, G.; Bashan, Y. (1996). Soil Biology and Biochemistry 28:1651–1660.

Jain, A.K. (2009). In: Plant Diseases Management for Sustainable Agriculture, Editor: Shahid Ahmad, Sheer-E-Kashmir University of Agricultural Sciences and Technology, (Regional Agricultural Research Station, Rajouri, Jammu, Published by Daya Publishing House, New Delhi pp. 299-327.

Jain, A.K.; Kumar, S.; Panwar, J.D.S. (2007). Advances of Plant Sciences India 20 (2): 337-339.

Jain, A.K.; Kumar, S.; Panwar, J.D.S. (2008). Legume Research 30 (3): 201-204.

Jimnez-Salgado, T.; Fuentes-Ramirez, L.E.; Tapia- Herandez, A.; Mascarua, M.A.; Martinez-Romero, E. and Caballero-Mellado, J. (1997). Applied and Environmental Microbiology 63: 3676–3683.

Kasiamdar, R.S.; Smith, S.E.; Smith, F.A. and Scott, E.S.: (2001). Plant and Soil. 238:235–244.

Katyal, J.C.; Venkatashwarlu, B.;Das, S.K. (1994). Fertiliser News 39(4): 27-32.

Kumar, G.N. (2003). Improvement of PGPR by transformation. Prospectus and Challenges. 6th International PGPR workshop held at IISR Calicut.

Kumara Swamy, C.A.; Raghunandan, B.L.; Chandrashekhar, M.; Brahmaprakash, G.P. (2010). Indian Journal of Science and Technology 3(7):689–692.

Loganathan. P.; Sunitha. R.; Parida. A.K.; Nair. S. (1999). Applied Microbiology 87:167–172.

Loganathan. P.; Nair. S. (2003). Biotechnology Letter. 25:497–50.

Lovelock, C.E.; Wright, S.F.; Clark, D.A.; Ruess, R.W. (2004). Journal of Ecology 92:278–28.

Mahdi, S.S.; Hassan, G.I.; Samoon S.A.; Rather, H.A.; Dar, Showkat A. and Zehra, B. (2010). Journal of Phytology 2(10):42–54.

McGonigle, T.P.; Miller, M.H. (1999). Applied Soil Ecology. 12: 41–50.

Meenakshisundaram, M.; Karrupagnaniar, S. (2010). International Journal of Biology and Medical Research. 1(4): 298–300.

Mosse, B., Stribley, D.P.; Le Tacon, F. (1981). Advance in Microbial Ecology 5:137-210.

Oliveira, R.G.B.; Drozdowicz, A. (1987). Zentralblatt für Mikrobiologie. 142: 387–391.

Olsson, P.A.; Thingstrup, I.; Jakobsen, I.; Baath, E. (1999). Soil Biology and Biochemistry 31:1879–1887.

Panwar, J.D.S.; Jain, A.K. (2015). Organic Farming- Scope and Use of Biofertilizers. New India Publishing Agency New Delhi.

Panwar, J.D.S.; Sirohi, G.S. (1989). Farmers and Parliament, 24:23-24.

Panwar, J.D.S.; Thakur, A.K. (1995). Proc: Mycorrhizae biofertilizers for the future. Tata Energy Res., Institute. pp. 13-15.

Ramarethinam, S.; Chandra, K. (2006). Psetology, 30(11): 35-39.

Rangrajan, S.; loganathan, P.; Saleena, L.M.; Nair, S. (2001). J. Appl. Microbiol. 91:742-749.

Rangrajan, Saleena, L.M.; Vasudevan, P.; Nair, S. (2003). Plant Soil. 251:73-82.

Rathi, B.K.; Jain, A.K.; Kumar, S. (2009). Journal of Indian Botanical Society 88(1&2): 1-3.

Rillig, M.C.; Wright, S.F.; Nichols, K.A.; Schmidt, W.F.; Torn, M.S. (2001). Plant and Soil. 233:167–177.

Roger, P. A.; Ladha, J.K. (1992). Plant Soil 141:41-5.

Saravanan, V.S.; Mahayana, M.; Osborne, Jabez.; Thangaraju, M.; Sa, T.M. (2008): Microboiol Ecology. 55:130–140.

Sarma, Y.R.; Rajan, P.P.; Paul, D.; Beena, N.; Anandaraj, M., (2000). In: Seminar on Biological Control with Pant Growth Promoting Rhizobacteria for sustainable Agriculture. University of Hyderabad, Hyderabad.

Sawers, J.H.; Caroline, G.; Paszkowski, U. (2007). A study about cereal mycorrhiza, an ancient symbiosis in modern Agriculture. Department of Plant Molecular Biology, University of Lausanne, Biophore Building, Lausanne, Witzerland.

Sevilla, M.; Kennedy, C. (2000). Colonization of Rice and other cereals by Acetobactor diazotrophicus, an endophyte of sugarcane. In The Quest for Nitrogen Fixation in Rice (Eds. J.K. Ladha and P.M. Reddy). International Rice Research Institute, Manila, Philippines, 15: 1-165.

Somers, E.; Ptacek, D.; Gysegom, P.; Srinivasan, M.; Vanderleyden, J. (2005). Applied and Environmental Microbiology. 71: 1803–1810.

Sudhakar, P.; Gangwar, S.K.; Satpathy, B.; Sahu, P.K.; Ghosh, J.K.; Saratchandra, B. (2000). Indian Journal of Sericulture. 39: 9–11.

Tapia-Hernández, A.; Mascarúa-Esparzá, M.A.; Caballero- Mellado, J. (1990). Microbios. 64: 73–83.

Tripathi, A.K..; Mishra, B.M.; Tripathi, P. (1998). J. Bio. Sci. 23: 463-47.

Yu, D.; Kennedy, I.R. (1995). Soil Biol. Biochem. 27: 459-462.

Zemen, A.M.M.; Tchan. Y.T.; Elmerieh, C.; Kennedy, I.R. (1992). Research in Microbiology Institute, Pasteur Elsevier 143: 847-855.

3

Liquid Biofertilizers

The earlier products of bio-fertilizers were carrier (solid) based where lignite is usually added as a carrier material. Lignite is hazardous to the production workers. Also, the shelf life of carrier based bio-fertilizers is only 6 months and is difficult to transport. Liquid Biofertilizers (LBFs) on the other hand have a shelf life of minimum one year, with no health hazards to production workers and are easy to transport. Additionally, LBF can be used in drip irrigation and as a component of organic farming. The liquid Bio fertilizers are suspensions having agriculturally useful microorganisms, which fix atmospheric nitrogen and solubilize insoluble phosphates and make it available for the plants. The use of this biofertilizer is environment friendly and gives uniform results for most of the agricultural crops and directly reduces the use of chemical fertilizer by 15 to 40%. The shelf life of the liquid bio fertilizer is higher (in the range of one to two years) compared to that of solid matrix base biofertilizer.

3.1 Definition

Liquid biofertilizers are liquid formulation containing the dormant form of desired microorganisms and their nutrients along with the substances that encourage formation of resting spores or cysts for longer shelf life and tolerance to adverse conditions. The dormant form on reaching the soil, germinate to produce fresh batch of active cells. These cells grow and multiply by utilizing the carbon source in the soil or from root exudates.

"A preparation comprising requirements to preserve organisms and deliver them to the target regions to improve their biological activity" or "A consortium of micro-organisms provided with suitable medium to keep up their viability for certain period which aids in enhancing the biological activity of the target site".

3.2 Classification of Liquid Biofertilizers

LBs are the microbial preparations containing specific beneficial microorganisms which are capable of fixing or solubilizing or mobilizing plant nutrients by their biological activity. Liquid formulation is a budding technology in India and has very specific characteristics and uniqueness in its production methods. They are broadly classified into three groups such as-

1) Nitrogen Fixing Microbes (NFM)
2) Phosphorus Solubilizing Microbes (PSM) and Phosphate Mobilizing Microbes (AM) and
3) Potash Mobilizing Microbes (*Frateuria aurentia*)

Among the three groups 'Phosphorus solubilizing microbes' usage is to the larger extent in the current agro market. The commercial production of Potash mobilizing bacteria has been developed recently. Solubilization or mobilization of micronutrients by exploiting the biological activity of microbes like as

1) Zinc and Sulphur Solubilizing Bacteria (*Thiobacillus* spp.)
2) Manganese solubilizer (*Pencillium citrinum*)

The biggest challenge of the biofertilizer production units is the viability of organisms up to field application. Some of the beneficial organisms are very effective *in vitro*, but may fail at some stage of cropping, even at the stage of harvesting. The common and potential causes of this demise are poor stability of the biofertilizer (by means of the viability) which may not be constant from prior storage to field application, very low potential ingredients reaching the active site of the plant practically and rapid degradation of the fertilizer by means of acclimatization problems. Liquid formulations of the beneficial organisms (LBFs) or bioinoculants are the promising inputs which can effectively combat the present agro adverse Scenario, resolving the low viability of the organisms, maximizing the efficacy of the inputs and satisfying the concept of cost-effectiveness. Anand Agricultural University (AAU), Anand, Gujarat is one of the universities that has developed a liquid formulation of bio-fertilizers that are safe and eco-friendly alternative to chemical fertilizers. LBFs are sold to farmers under the brand name "Anubhav liquid Bio-fertilizers" by the University (Fig. 3.1). Anubhav LBF is based on native cultures of bacteria viz., *Azotobacter chroococcum*, *Azospirillum lipoferum* and *Bacillus coagulans*. To extend its reach to the farmers, the AAU has licensed the technology of LBF for commercialization to three companies in Gujarat through its Business Planning and Development Unit (BPDU) under Public Private Partnership (PPP) mode. BPDU is a special project at AAU, Anand under the World Bank funded scheme of National Agricultural Innovation Project (NAIP), Indian Council of Agricultural Research (ICAR), New Delhi. AAU has supplied LBFs to the tune of 50,000litres to the Government of Gujarat for distribution to farmers as a part of Krishi kit during Krishi Mahotsav, a mass agricultural technology dissemination programme of the Department of Agriculture, Government of

Gujarat. The response of Gujarat farmers on use of LBF in different crops such as cotton, banana, potato, rose, turmeric, papaya etc. reported better yield and quality M/s Kumar Krishi Mitra Bio Products Ltd., Pune, has already commenced production. There is a growing demand for organic foods in the global market. The use of these liquid bio-fertilizers would help the Indian farmers to produce organic crops so as to compete in the global market.

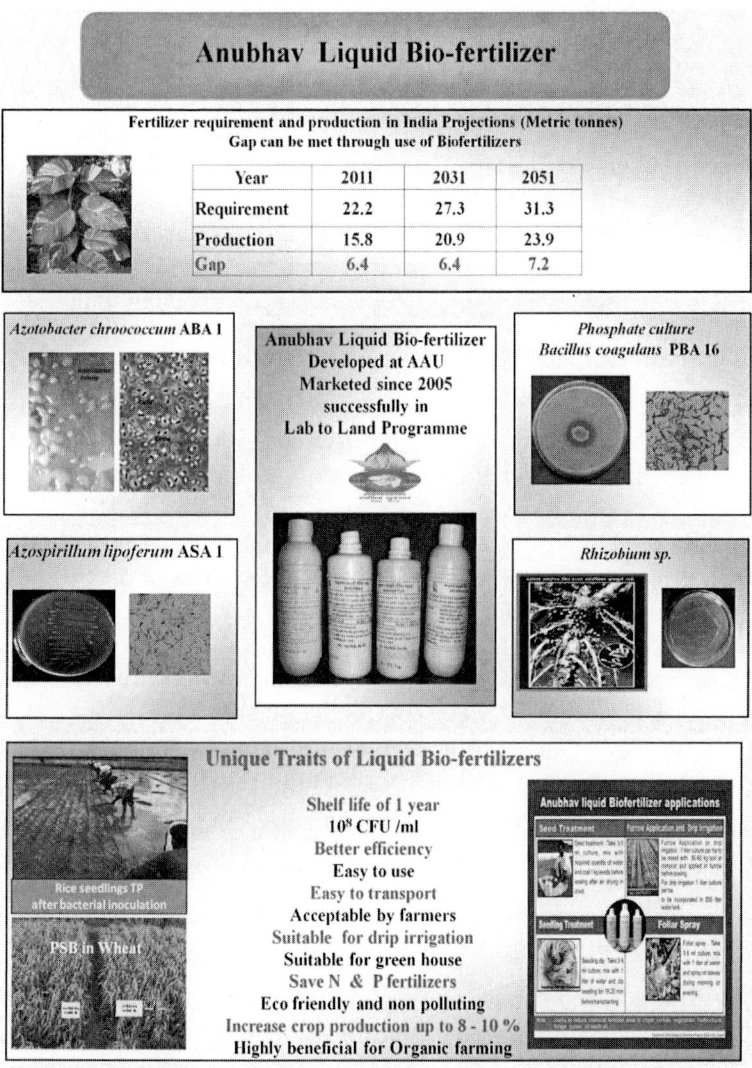

Fig. 3.1: Anubhav Liquid Biofertilizer (*Source:* http://www.icar.org.in/en/node/5667 Technology for Liquid Bio-Fertilizers commercialized)

3.3 Types of Inoculums

A wider range of formulations with additives are also available in the global agri market. Basically liquid biofertilizers have been classified as:

3.3.1 Dry Inoculum Products

These formulations comprise inoculums in semi dry condition. This includes dusts, granules and briquettes, based on particle or aggregate size. This group also comprises wettable powders formulated in dry powder. These powders require liquid or water as carrier just before the application to activate the inoculums. Dusts usually ranges from 5-20 mm in size and require low absorbance inert diluents as carriers. Low dusts particles, i.e. less than 10 mm size adhere best on the target site but are hazardous when inhaled. About 30% of the dry weight of dusts contains organisms as suspension. They are normally prepared by feeding the organism into an air stream for mixing with a blender. Particle size, bulk density and flow ability are extremely important.

3.3.2 Granules, Pellets, Capsules and Briquettes Inoculum

These include masses of different sizes which are usually called as granules, pellets, capsules and briquettes. Granules are discrete masses 5-10 mm^3 size, pellets are of size >10mm^3, and briquettes are large blocks up to several cubic centimeters; just like dusts, these products also contain inert carriers such as clay minerals, starch polymers and ground plant residues holding the organisms. Absorption (more important for formulating slurries of organisms), hardness, bulk density and product disburse rate in water etc., are the major factors taken into consideration in selecting a suitable carrier. Soft carriers e.g. Bentonite, disburse quickly to release the organism. Various materials are used to coat the product to slow down or control the rate of release, which also varies with the unit size. The usual concentration of organisms in granules is 20- 30%. There are three types of granular formulations in general:

- The organisms attached to the outer surface of a granular carrier by a sticker in a rotating drum.
- The organisms incorporated into a paste or powdered carrier which sets as a matrix size being controlled by passing the product through a sieve.
- The organisms sprayed onto a rotating granular carrier without a sticker. When the carrier forms a protective coat around a core aggregate of organisms, the unit is termed as a capsule.

3.3.3 Wettable Powder Inoculums

These formulations comprise charcoal, lignite, vermiculite powders which are predominated among all commercial products. These are blended with 3% gum to make them stable on the shelf during storage and stick readily to the seeds.

3.4 Issue of Formulations

Formulation comprises aids to preserving organisms and to deliver them to their target fields and – once there-to improving their activities. A technical concentrate of an organism that has been achieved by a particular process is called as a formulation, or a product, which may be stored and put on sale commercially. A product often does not fully serve all the requirements of organisms up to the field. There are varieties of formulation both in liquid and solid. The main types currently used for biofertilizers have been classified into dry products (dusts, granules and briquettes) and suspensions (oil or water-based and emulsions). A wider range of formulations with additives are available in market. Liquid formulations are sprayable propagules of the microbial agents suspended in suitable liquid medium (or) A consortium of microorganisms provided with suitable medium to keep up their viability for certain period which aids in enhancing the biological activity of the target site. Basic characteristics of formulation is given below:

- Enhance activity of the organism at the target site by increasing its activity, reproduction, contact and interaction with the target crops.
- Easily delivered to the field in the most appropriate manner.
- Protect the micro-organism from harmful environmental factors at the target site (field), thereby increasing persistence.
- Stabilize the organism during production, distribution and storage.

LBFs are special liquid formulation containing not only the desired beneficial microorganisms and their biological secretions, but also special cell protestants or substances that encourage the formation of dormant spores or cysts for longer shelf life and tolerance to adverse conditions. Although, India boasts to be the largest biofertilizers producer in the globe the creditability of biofertilizers is slow among farmers. Apart from the carrier based inoculums, considering above said limitations in carrier based Regional Centre of Organic Farming (RCOF), Bangalore formulated the liquid based biofertilizers in year 1998 for all types of biofertilizers and biopesticides. This technology is an alternative solution to carrier based biofertilizers, biopesticides and bio-control agents and this is not the usual broth culture from the fermenter or a water suspension of the carrier based biofertilizers.

3.4.1 Dormant Aqueous Suspensions

Several current commercial products available in market follow the dormant technology. Generally, they are using growth suppressants, contaminant suppressant like Sodium azide ((NaN_3), Sodium benzoate ($C_7H_5NaO_2$), Butanol (C_4H_9OH), Acetone ($(CH_3)_2CO$), Fungicides, Insecticides etc. for the long term viability, while these formulation in particular is not suitable for organisms whose fast actions are required immediately for short duration crops. It was noted the *Rhizobium* dormant liquid formulations when used after 8 months reduce the size of nodules and effect N_2 fixation process due to the need of long duration of activation time. In case of *Azospirillum, Azotobacter*, PSM, it was observed that these bacteria crossed extreme dormant stage. .

3.4.2 Dormant Oil Suspension

Micro-organisms can be suspended in oil at high concentration in various degrees of dehydration and remain viable. This formulation delivers organisms in a physiologically dormant state and does not encourage the growth of contaminants during storage. The bacteria/ fungus has been successfully dried by continuous aeration as a suspension in oil to provide inoculants with shelf life of several years. Oil suspension also has the limitation in reactivating the bacteria/fungus. It generally takes 10 to 20 days to regenerate and the initial requirement of the plant is not fulfilled by them. The liquid suspension contains particular microorganisms about 40%, suspender ingredient 1-3%, dispersant 1-5%, surfactant 3-8% and carrier liquid (water) 35-65% by weight. Viscosity roughly equals the setting rate of the particles. This is achieved by the use of colloidal clays, polysaccharide gums, starch, cellulose or synthetic polymers.

3.4.3 Biosurfactant for Improving Liquid Formulation

For improving technology in liquid formulation, efforts are being made to add Biosurfactant. These are surface active biomolecules produced by a variety of microorganisms. These amphiphilic are generally produced on microbial cell surfaces but also through extracellular secretion (Muthuswamy *et al.,* 2008). These surfactants are used as emulsifier and are safer than chemical emulsifiers. The common biosurfactants are Glycolipids, Lipopeptide Rhamnolipids and Sophorolipids (Table 3.1). There is enough scope to use such biosurfactants for the development of liquid biofertilizers.

Table 3.1: Some Important Biosurfactants for Improving the Quality of Liquid Formulation

Sl No.	Type of biosurfactants	Produced by	Substrate used
1	Glycolipids	*Candida apicola*	Oil refinery waste
2	Lipopeptide	*Serratia marcescens*	Sunflower oil
3	Rhamnolipids	*Pseudomonas aeruginosa*	Curd whey and distillery waste
4	Sophorolipids	*Candida lipolytica*	Babassu oil waste

3.5 Factors Affecting the Liquid Formulation

3.5.1 Temperature

Temperature is important factor for the shelf life of microbial products and it can affect their activity before or after application. Colonization proceeds at field temperatures in the cropping season, but slows at lower temperature. Strains used in liquid formulation normally grow at 37^0C, (Chandra *et al.,* 1999) and able to tolerate temperature up to 45^0C for two year or more. Whereas, solid based shelf life is hardly up to 3 months as rise in temperature beyond 35^0C and start rapid decline of organisms.

3.5.2 Acclimatization

Effect of environment on bio fertilizers has been studied by many workers. It is reported that it has negative effect on different strains, but all the research was done on the carrier based biofertilizers. It is observed that the efficiency of liquid is almost same in all environments, but efficiency may reduce 20-25% in different climatic conditions in case of solid base. Temperature, humidity, leaf surface exudates and competitors, etc are the critical factors which inactivates the microbes. They may lost physically from the target location by the action of wind, rain or leaching depending on the fact why and where the product is used.

3.5.3 Effect of Humectants

The effect of moisture content on storage stability, some organisms may need moisture for its activity this need is fulfilled by liquid inoculum but in case of carrier based inoculum bacteria get stressed, when carrier become dry during transport and storage. Bacteria used for plant growth need the plant surface to be wet in order to establish them. These needs can be overcome by only liquid formulation as product contains humectant. In general, there is little direct effect of relative humidity on the spore forming bacteria in liquid form.

3.5.4 Sunlight Intensity

Generally microbes are sensitive to temperature. The most harmful rays UVB (280-320nm) and UVA (320-400nm) reaching the earth's surface have direct effect on microbes. Of course some are less damaging where as some have the maximum effects. However, microbes are sensitive to the wave-lengths outside this range. To counter the harmful effects of high temperature sunscreens are added to a formulation. Sunscreens act by reflecting and scattering physically or by absorbing radiation selectively, converting short wavelengths to harmless longer ones. However, no such types of sunscreens are available in solid base to resist the effect of sunlight.

3.5.5 pH Value

The pH of a product plays a vital role in liquid inoculum preparations. It must be stabilized within certain ranges. Very low and high pH conditions will normally inactivate agents. A buffer therefore is maintained by adding some additives which render the better shelf life in liquid. Maintenance of optimal pH improve shelf-life of some of microorganisms like *Azotobacter, Azospirillum*, Phosphorus Solubilizing bacteria (PSM), Potash mobilizing bacteria (KMB) *Frateuria aurentia* (Chandra *et al.,* 1999).

3.5.6 Plant Exudates

Plant exudates may influence the efficiency of liquid biofertilizer in various ways. These include chemicals present in the soil, as well as plant secretions and chemicals on the phylloplane. For example, on cotton and many other xerophytic plants, high levels of magnesium (Mg), calcium (Ca) and manganese (Mn) as carbonates and bicarbonates are exuded onto the leaf surface; this can raise the pH of the leaf surface to values as high as 10 or 11. The presence of these ions or the associated pH value, when wetted by dew or water spray, can inhibit or inactivate some organisms. To overcome with these problems only liquid proper formulations take care as it has a buffering property which neutralize the effect of temporary pH effect. However, Potash Mobilizing bacteria (KMB) *Frateuria aurentia* has no effect as it can tolerate even at pH 11.

3.5.7 Stabilization of Liquid Inoculum

The longest period of time in the life of a product elapses during storage, i.e. the time between manufacture and eventual use in the field. This period can range from several weeks to years. Organisms are required to remain viable during storage, with minimum loss of potency/ activity and without loss or breakdown of the desired formulation properties. The authors treat the liquid formulations for improving stability by appropriate growth conditions during production by appropriate storage prior to formulation by appropriate processing after production. In addition, or alternatively, additives are included to improve stability. Depending on the organisms involved, different problems must be addressed to improve stability. It is suggested during preparation of base of liquid some additives that inhibit germination of the organisms after application must be avoided.

3.5.8 Application of Liquid Inoculum

Addition of surfactant or oil and emulsifier to water, or use of pure oil, forms drops of more even size than those of water alone, with consequently better controlled spray. Oil is preferred for ultra-low-volume (ULV) sprays, water is normally used as the diluent at higher volumes. Liquid formulations contain high count ranges from 10^9 to 10^{10} cells per ml which optimize spray droplet size to maximum coverage of the target and ensures maximum number of effective bacteria for action.

3.5.9 Handling of Liquid Inoculums

Liquid inoculum ensures that the product is easy to handle and apply. For example, in suspensions thickeners or suspenders added by the authors which help in maintaining even distribution of the organism. The liquid prevent clumping of the micro-organism and ensure its ready re-suspension after prolonged storage. Dusts and wettable powders are not able to maintain uniformity. It is shown that in case of *Azospirillum* the population come down up to 10^5 at six months duration at room temperature where as in liquid survives up to 2 years and population maintain up to 10^8/ml. Similarly in *Azotobacter*, KMB, *Rhizobium* solid maintain the shelf life up to 6 months except PSM which services up to 8 months but in the case of liquid formulations in *Azotobacter*, PSM and KMB survives up to 2 years followed by *Rhizobium* only 14 months. This shows the superiority of liquid formulation over carrier base formulations.

3.6 Advantages of Liquid Biofertilizers

The advantages of liquid biofertilizer over conventional carrier based biofertilizers are given below:

- Better survival on seeds and soil and easy to use by the farmers.
- Cost saving on carrier material, pulverization, neutralization, sterilization, packing and transport.
- Dosages are 10 times less than carrier biofertilizer.
- High commercial revenues and export potential.

- High populations can be maintained more than 10^9 cells/ml up to 12 months to 24 months.
- Identified by typical fermented smell.
- Longer shelf life as 12- 24 months.
- No loss of properties due to storage upto 45^0C since no effect of high temperature.
- No need of running biofertilizer production units throughout the year.
- Quality control protocols are easy and quick.
- Very high enzymatic activity since contamination is nil.

Further, it is very necessary to understand that, "liquid biofertilizers are not merely usual broth cultures from fermenters which are specially prepared with desired agriculturally important micro-organisms along with proper nutritional base to facilitate protection of organisms from adverse conditions and resting spore formation or cysts etc. Many researchers still have doubts about the liquid formulations. Although, in commercial all bacterial inoculum being well adopted by the users

3.7 *Rhizobium* as Liquid Biofertilizers

It belongs to bacterial group and the classical example of symbiotic nitrogen fixation. The bacteria intact the legume root and form root nodules within which they reduce molecular nitrogen to ammonia which is readily utilized by the plant to produce valuable proteins, vitamins and other nitrogen containing compounds. The site of symbiosis is within the root nodules. It has been estimated that 40-250kg N/ha/year could be fixed by different legume crops by the microbial activities of *Rhizobium* (Table 3.2). It is dull white in colour and has no bad smell and foam formation and requires pH 6.8 to 7.5.

Table 3.2: Quantity of Biological N fixed by Liquid Rhizobium in different Crops

Sl. No.	Host Group	*Rhizobium* Species	Crops	N fix kg/ha
1.	Pea Group	*Rhizobium leguminosarum*	Green Pea, Lentil	62-132
2	Soybean group	*R. japonicum*	Soybean	57-105
3	Lupini group	*R. lupini orinthopus*	Lupinus	70-90
4	Alfalfa group	*Rhizobium meliloti, Medicago rigonella*	Melilotus	100-150
5	Beans group	*Rhizobium. phaseoli*	Phaseoli	80-110
6	Cicer group	*Rhizobium sp.*	Bengal gram	75-117
7	Clover group	*Rhizobium trifolii*	Trifolium	1 3 0
8	Cowpea group	*Rhizobium* sp.	Moong, Gram, Cowpea, Ground nut	57-105

Source: Chandra *et al.,* (2005)

Use of Fungicides/Pesticides/Herbicides with Liquid *Rhizobium* inoculum

The effect of chemicals on nodulation and nitrogen fixing *Rhizobium* is negligible. Seed dressing fungicides such as Thiram and mercury have some adverse effect on the viability of *Rhizobium*, which can be overcome by doubling the doze of *Rhizobium* culture. Studies suggest that *Rhizobium* strain MRL3 may be exploited as a bioinoculant to augment the efficiency of lentil exposed to insecticide-stressed soils (Ahmed and Khan, 2011). On the other hand fungicides such as Dithane M-45 and Bevistin do not have any adverse effect on *Rhizobium* and the minimum tolerance doses with different fungicides, corbandazium, Dithane M-45,

Use of Chemical Fertilizer along with Liquid *Rhizobium*

Most pulses and legume oil seed crops require 30-40kg Nitrogen and 40kg Phosphorus/ha for better growth and yield. If crop is provided chemical nitrogen only 5-10kg/ha as a basal dose; is essential to fulfill the initial nitrogen requirement of crop up to 20-25 day i.e. till the *Rhizobium* nodules start their functioning. Phosphorus is essential for better *Rhizobium* inoculation so 30-40kg of phosphorus/ha is required as basal dose, if PSM is used the dose of chemical Phosphorus can be 15-20kg/ha. Potassium is also essential for better *Rhizobium* inoculation so 30-40kg of Potassium/ha is required as basal dose, if KMB is used the dose of Potassium chemical can be 10-15kg/ha.

Use of Sticky Materials for Seed Treatment

Generally sticky materials are used for the easy adhesion of the inoculums onto the target plant parts and subsequently increase bacterial population on the seeds for maximum nitrogen fixation. The germination and nodulation tests were performed on the pelleted seeds and results showed that although the sugar and molasses initially bound less inoculant and lime to the seed, the number of surviving *Rhizobia* was similar to that obtained by the Gum Arabic treatment after storage at 27.50°C for 5 days for fast growing *Rhizobia*. The inoculated pelleted seeds of slow growing *rhizobia* showed promising results in germination and nodulation in acidic soil.

3.8 Azospirillum as Liquid Biofertilizers

Azospirillum belongs to bacteria and is known to fix considerable quantity of nitrogen in range of 20-40kg N/ha in the rhizosphere in non-leguminous plants such as cereals, millets, oilseeds, cotton etc. *Azospirillum* is considered as the efficient biofertilizers because of its ability of inducing abundant roots in several plants like rice, millets and oilseeds even in upland conditions.

An estimated amount of 25-30% chemical nitrogen fertilizer can be saved by the appropriate use of *Azospirillum* inoculants. The genus *Azospirillum* has three species viz. *A. lipoferum, A. brasilense* and *A. amazonense*. These species have been commercially exploited for the use of nitrogen supplying biofertilizer. One of the characteristics of *Azospirillum* is its ability to reduce nitrate and denitrify. Table 3.3 presents occurrence of *Azospirillum* and their nitrogen fixing capacity in different plants. Both the species *A. lipoferum* and *A. brasilense* may comprised of strains which may denitrify actively or weakly or reduce nitrate to nitrite. Therefore, for inoculation preparation, it would be necessary to select strains which do not possess these characteristics. *Azospirillum lipoferum* present in the roots of some tropical forage grasses such as Digitaria, Panicum, Brachiaria, Maize, Sorghum, Wheat and Rye (Pindi and Satyanarayana, 2012; Gautam *et al.,* 2017).

Table 3.3: Occurrence of Azospirillum and N_2 fixing capacity bacteria in the roots of several plants and the amount of nitrogen fixation

Plant	g N_2 fixed/g of substrate
Cynodon dactylon	36
Paddy (*Oryza sativa*)	28
Panicum Sp	24
Sorghum (*Sorghum bicolor*)	20
Maize (*Zea mays*)	20
Amaranthus spinosa	16
Setaria Sp.	12

Source: Chandra *et al.,* (2005)

Production of Growth Hormones

Azospirillum cultures synthesize considerable amount of biologically active substances like vitamins, nicotinic acid, indole acetic acid, gibberellins. All these hormones/ chemicals help plants in better germination, early emergence, and better root development.

Field Applications of Liquid *Azospirillum*

- Aids utilization of potash, phosphorus and other nutrients
- Encourages plumpness and succulence of fruits and grains and increases protein percentage.
- Stimulates growth and produces green color characteristics of a healthy plant.

Inoculants of *Azospirillum* have been proved beneficial in a variety of crops including cereals, forage and other crops. The crop response however, varies greatly with types of crop and variety, location, season, soil fertility level; native microorganism interaction etc., the ability to fix atmospheric nitrogen and to produce phytohormones determines the crop response and yield. A series of experiments have been conducted on a range of crop varieties to evaluate the crop response to *Azospirillum* inoculants and a wide variation in production up to 40% have been observed. Non-functioning of *Azospirillum* in the fields can be identified by yellowish green color of leaves which indicates very poor or no nitrogen fixation which in turn results in arresting the growth promotion

3.9 *Azotobacter* as Liquid Biofertilizers

Azotobacter belongs to the genus diazotrophic bacteria which is a free-living organism whose resting stage is a cyst. It is abundantly found in neutral to alkaline soils, in aquatic environments, and on some plants. *Azotobacter* is capable of performing several metabolic activities, including atmospheric nitrogen fixation by conversion to ammonia. *Azotobacter* spp. has the highest metabolic rate of any organisms. It serves as potential biofertilizer for all non-leguminous plants especially rice, cotton, vegetables etc. *Azotobacter* population is high in rhizosphere region and the rhizoplane does not contain *Azotobacter* cells. The lack of organic matter in the soil is a limiting factor in the proliferation of *Azotobacter* in the soil.

Liquid *Azotobacter* in Tissue Culture

- The performance of *Azotobacter* liquid inoculants was comparatively better than all the treatments in 10% MS medium followed by *Azospirillum*.
- The performance of *Azotobacter* liquid inoculants was comparatively better than all the treatments followed by *Azospirillum* for the growth of polybag sugarcane seedlings.

Liquid *Azotobacter* as Bio-control Agent

Azotobacter have been found to produce some effective antifungal substance which inhibits the growth of some soil fungi like *Aspergillus, Fusarium, Curvularia, Alternaria, Helminthosporium,* and *Fusarium* etc.

3.10 *Acetobacter* as Liquid Biofertilizers

Acetobacter (a sacharophilic bacteria) associated with sugarcane, sweet potato and sweet sorghum, coffee plants is acid-producing bacteria which fix atmospheric nitrogen. In fact, the *A. diazotrophicus*-sugarcane relationship is beneficial symbiotic relationship between grasses and bacteria which fixes about 30kg N/ha/year of nitrogen. This bacterium is being exploited and commercialized extensively for sugarcane crop reducing the primary dependency in the chemical nitrogen fertilizer. It is known to increase cane yield 10-20 ton/acre and sugar content by 10-15%.

Acetobacter pasteurianus for Sulphur

The physical structure of sulphur consists of several amino acids, including methionine and cystine, which are essential components of plant and animal proteins. Sulphur is important in synthesis of nitrogen in plants; sulphur acts more like nitrogen than any other essential plant nutrients. Indeed, sulphur deficiency symptoms in plants closely resemble nitrogen deficiency symptoms. Moreover, leaching losses of sulphur from the soil are very much like those of nitrogen caused by percolation waters through light-textured sandy and salty soils. Thus, there is need for annual supplies of sulphur to meet the requirements of growing plants. Although, we are accustomed to use sulphur in the form of fertilizers but number of such farmers are very less, due to its cost factors etc. It was estimated that about 70-90% of total surface soil sulfur is found in organic matters. The remainder of the sulphur is present as sulfides and as soluble and insoluble sulfates. Usually, only 5-10% of the sulfates are found in the surface furrow slice of humid region, cultivated soils. In humid region soils, most of the sulfates are of 12-14 inches deep and are associated with oxides of iron and aluminum, especially in acid soils having kalononitic type clay. These bacteria help in converting this non-usable form to usable form. It was experienced that use of bacteria 200 ml/acre influenced in level of sulfur in crops like vegetable, cabbage, turnip, onion, cotton, fruits, oilseeds, spices etc.

Liquid *Acetobacter diazotrophicus* in Sugarcane

Application of *Azospirillum* and Phosphobacteria for the cash crop sugarcane has been a regular practice for the past few years which is saving nearly 20% chemical nitrogen and phosphate applications in South India. It is evident from various studies that *Acetobacter diazotrophicus* present in sugarcane stem, leaves, soils have a capacity to fix up to 300kg of nitrogen. This bacterium first reported in Brazil where sugarcane is cultivated in very poor sub-soil fields which are fertilized with phosphate, potassium and micro elements alone could produce a yield for three consecutive recorded harvests of 182 to 244 tons per hectare, without the use of any chemical nitrogen fertilizer. This manifests the assumption that active nitrogen fixing bacteria must be associated within the plant regularly. An extensive research work has been conducted by the scientists of Brazil and Australia on bacteria like *Acetobacter diazotrophicus, Herbaspirillum seropedicae* and *Herbaspirillum rubisubalbicans* in leaves, stems and roots and revealed the fact that they are largely responsible for atmospheric nitrogen N-fixation.

3.11 Phosphorus Solubilizing Micro-organisms (PSMs)

Microbial mineralization, solubilization and mobilization are the most important aspects of the phosphorus cycle, besides chemical fixation of phosphorus in soil. The enzymatic activity of microorganisms is responsible for the mineralization of organic phosphorus which is left over in the soil after harvesting, or added as plant or animal residues to soil (Bot and Bonetis, 2005; Eckert, 2012). Phosphorus solubilizing bacteria and fungi play a vital role in persuading the insoluble phosphatic compound such as rock phosphate, bone meal and basic slag and particularly the chemically fixed soil phosphorus into available form. These special types of microorganisms are known as Phosphate Solubilizing Microorganisms (PSM) which includes different groups of microorganisms such as bacteria and fungi that convert insoluble phosphatic compounds and fixed chemical fertilizers into soluble form. The species of *Pseudomonas, Micrococcus, Bacillus, Flavobacterium, Penicillium, Fusarium, Sclerotium, Aspergillus* and some other are considered as active in biophosphorus conversion.

These bacteria and fungi can be grown in the media containing $Ca_3(PO_4)_2$, $FePO_4$, $AlPO_4$, apatite, bone meal, rock phosphate or similar insoluble phosphate compounds are the sole source of phosphate. Such organisms not only assimilate phosphorus but also trigger the release of an excess amount of soluble phosphate than their actual requirements. Several rock phosphate dissolving bacteria, fungi, yeasts and *actinomycetes* were isolated from soil samples collected from rock phosphate deposits and rhizosphere soils of different legume crops. Some of the isolates solubilized and made available phosphorus to crop from rock phosphate and many isolates solubilized a very high quantity of tricalcium phosphate. The most efficient bacterial isolates were identified as *Pseudomonas striata, Pseudomonas rathonis, Bacillus polymyxa, B. megatherium.* Apart from these, fungal isolates such as *Aspergillus awamori, Penicillium digitatum, Aspergillus niger, Schwanniomyces occidentalis* are identified as the potential phosphorous solubilizers. These efficient microorganisms have shown consistency in their capability to solubilize chemically fixed soil phosphorus and rock phosphate from different sources.

Effect of Liquid PSM under Field Conditions

- Encourages early root development.
- Helps rapid cell development in plants and consequently increases resistance towards diseases.
- Increase micro nutrients in soil like Mn, Mg, Fe, Mo, B, Zn and Cu etc. depending upon the nutrient presents in non-available form.
- Increases the compatibility of other beneficial microbes with plants.
- PSM produced organic acids like malic, succinic, fumaric, citric, tartaric and alpha ketoglutaric acid which hastens the maturity and thereby increases the ratio of grain to straw as well as the total yield.
- Stimulates formation of fats, convertible starches and healthy seeds.

Inoculants of phosphate solubilizing bacteria as fertilizer increases P uptake by the plant and enhance crop yield. There are many strains of different genera which promote the phosphate solubilization among which, strains from the genera *Pseudomonas, Bacillus* and *Rhizobium* are the most powerful phosphate solubilizers. Phosphate solubilization by the organisms mainly involves the production of organic acids which is the principal mechanism for solubilization, and the organic phosphorous in the soil is mineralized by acid phosphates which is a vital step. Cloning of several phosphatase encoding genes which are isolated is done and characterized for the better phosphate solubilization. Genetic manipulation of phosphate-solubilizing bacteria followed by their expression in selected rhizobacterial strains is the interesting approach among the scientific community for the optimum phosphate solubilization (Rodríguez and Fraga, 1999). Ghosh (2014) reported the reducing dependence on chemical fertilizers and its financial implications for farmers in India

3.12 Phosphate Mobilizing Micro-organisms (PMMs)

Compared with the other major nutrients, phosphorous is the least mobile and available to plants in most of the soils. In spite of its abundance in the most of the soils both in organic and inorganic forms, phosphorus is prime limiting factor for plant growth. Moreover plant species, nutritional status of the soil and ambient soil conditions determine the bioavailability of soil inorganic phosphorus. To circumvent the phosphorus deficiency to the soil phosphate solubilizing microorganisms (PSM) could play an important role in supplying phosphate to plants in a more suitable and eco-friendly manner (Khan *et al.*, 2009).

The later one endomycorrhizae are known as Arbuscular mycorrhiza which possess special structures known as vesicles and arbuscules, the latter helping in the transfer of nutrients from soil into root system. These fungi are classified on the basis of their spore morphology in seven genera namely Glomus, Gigaspora, Acaulospora, Sclerocystis, Scutellospora, Entrophospara, Archeospora as para Glomus. In general, two typical types of mycorrhizal fungi are found in associations with plant roots namely as *ectomycorrhizae*, *endomycorrhiza* and *ectendomycorrhiza*. Mycorrhizae increase the surface area of the plant root system for better absorption of nutrients from soil especially in phosphorus deficient soils. The *endomycorrhiza*, also known as *Arbuscular mycorrhiza*, aids in transfer of nutrients from soil into root system through specialized structures known as vesicles and arbuscules. The potential role of *mycorrhiza* in mobilizing & phosphate uptake and plant growth is being widely recognized in the recent times and *Arbuscular mycorrhizae* (AM) are the most common type which is being highly exploited. Advancement in biotechnology has promoted experimentation in artificial inoculation of plants with AM. Trials have mainly involved broadcasting and raking-in inocula over plant-growing substrates rather than application of inocula to seed before sowing. Since large-scale production of AM in axenic culture is not yet attained, inocula have been produced in pot cultures or in small field plots on plants grown under carefully controlled conditions, to avoid contamination by plant pathogens. Such inocula have comprised of infected roots or spores and hyphae trapped in soil, peat and clay carriers. Introduction of AM into field on a large scale has tended to use inoculum at excessively high and impractical rates in order to guarantee rapid infection by the introduced fungi. Application of inoculum directly to seed may reduce the amounts of inoculums required. Experimental inoculation has involved coating spores, mycelium and infected root fragments, wet-sieved from soil, onto seeds using methyl cellulose as sticker.

According to the studies of root nodulation as well as N and P, uptake is improved by green gram plant on the application of favorably interacting rhizospheric microorganisms as the inoculants and hence, yield is increased in the phosphorous deficient soils (Zaida *et al.,* 2004) *Bradyrhizobium* sp. (*Vigna*) and phosphate solubilizing microorganisms (PSM), *Pseudomonas striata* or *Penicillium* variable have shown a significant effect in enhancing the nutrient uptake and plant yield. Combined inoculation treatment of AM fungus, *Glomus fasciculatum* with *Bradyrhizobium* sp. (*Vigna*) + *P. striata* further augmented the nutrient uptake and plant yield (Khan *et al.,* 2009).

Benefits and Constraints

As the fungal hyphae facilitate increased surface area of absorption of nutrients like P, Zn, Cu etc., up to an extent of 8 cm the biochemical reactions in plant cycle are not interrupted promoting the plant growth. Moreover, it is rather easy to inoculate on crops which are normally grown on nursery beds or root trainers or poly bags because of the difficulty in mass producing AM fungi. Nearly 25 to 50% of phosphatic fertilizers can be saved on a year with the inoculation of efficient AM fungi. In spite of many advantages, AMF have some constraints which causes their limited usage. As it is an obligate symbiotic organism it need a host to survive, and it cannot be cultured in synthetic media in laboratory for large scale production. Variations and specifications in plant genotypes and soil nature make it difficult to use appropriate *mycorrhizal* strains for the phosphorous mobilization. Slow growth and development of *mycorrhizal* fungi association with the plant roots and poor resistance to the fungicides are some of the other potential constraints which are to be overcome by the rigorous field research with different approaches and optimizations.

3.13 Potash Mobilizing Bacteria (KMB) as Liquid Biofertilizers

In addition to the regular biofertilizers like *Rhizobium, Azospirillum, Azotobacter, Acetobacter* and Phosphate solubilizing/mobilizing and microbes to meet N&P nutrition in plants, microbes responsible for Potassium mobilization has been isolated and developed from Banana rhizosphere which is having the ability to mobilize the elementary or mixture of potassium which can be easily absorbed by plants. It is estimated that 50-60% of potash chemical fertilizers usage can be reduced by using *Frateuria aurentia*, a new bacterial species (species conformation by IMTECH (Chandigarh) as a bioinoculant. This new bacteria belongs to the family *Pseudomonaceae* have the extra ability to mobilize K in almost all types of soils especially, low K content soils, soils of pH 5-11 and it survives in the temperature up to 42°C. This potash mobilizing biofertilizers can be applied in combination with *Rhizobium, Azospirillum, Azotobacter, Acetobacter*, PSM etc. Potash mobilizing bacterial based product containing *Frateuria aurentia* producing plant growth promoting substances which offers plant a multifaceted benefits in terms of growth, by mobilizing potash and making it available to crops. It also enhances the efficiency of chemical fertilizer (Patel, 2011). Since large-scale production of AM in axenic culture is not yet attained, inocula have been produced in pot cultures or in small field plots on plants grown under carefully controlled conditions, to avoid contamination by plant pathogens. Such inocula is comprised of infected roots or spores and hyphae trapped in soil, peat and clay carriers. Introduction of AM into field on a large scale has tended to use inoculum at excessively high and impractical rates in order to guarantee rapid infection by the introduced fungi. Application of inocula directly to seed may reduce the amounts of inocula required. Experimental inoculation has involved coating spores, mycelium and infected root fragments, wet-sieved from soil, onto seeds using methyl cellulose as sticker. Optimization studies on the liquid potassium mobilize revealed following facts. Arbuscular mycorrhizae (soil base) and Potassium mobilizing bacteria, individually, as well as in combination with organic compost and phosphocompost shown better results in capsicum, Hungarian yellow variety when compared to the controls (plain soil). Potassium mobilizing bacteria (*Frateuria aurentia*) enriched phosphocompost shows the highest results in growth parameters next to all the biofertilizers applied together. This may be because of significant increase in mobilization and uptake of K leading to release of growth hormones by the K-mobilizer.

Advantages of AMF Application in Micropropagated Plants

- Better survival and establishment.
- Corrects the physiological defects of tissue culture plantlets.
- Enhances the absorption and transport of water and nutrients.
- Healthy and vigorous plantlets.
- Hormonal effects on shoot and root development.
- Reduces the hardening period.
- Resistance to pathogenic infection.

Effect of Liquid Potassium Mobilizer with AM

Optimization studies on the liquid potassium mobilize revealed following facts. *Arbuscular mycorrhizae* (soil base) and potassium mobilizing bacteria, individually, as well as in combination with organic compost and phosphocompost shown better results in capsicum, *Hungarian yellow* variety when compared to the controls (plain soil). Potassium mobilizing bacteria (*Frateuria aurentia*) enriched phosphocompost shows the highest results in growth parameters next to all the biofertilizers applied together. This may be because of significant increase in mobilization and uptake of K leading to release of growth hormones by the K-mobilizer. Improves plant growth is attributed to better uptake of nutrients like P, Zn, Cu etc. This is mainly because of increased surface area of absorption of nutrients by the hyphae. Mycorrhizal plant roots can extend up to 8cm. Because of the difficulty in mass producing AM fungi, the best way to utilize AM fungi for crop production would be to concentrate on crops which are normally grown on nursery beds or root trainers or poly bags, where they could easily be inoculated with desired AM fungi and transplanted to the field. Nearly 25-50% of phosphatic fertilizers can be saved through inoculation with efficient AM fungi.

3.14 Zinc and Sulphur Solubilizing Bacteria

Thiobacilus ferrooxidans also known as *Acidithiobacillus ferrooxidans* and *Thiobacillus thioxidans* and *Thiobacilus acidophilus* is an acidophilic bacteria present in water or soil. The bacteria produce number of organic acids. They are able to produce and tolerate highly acidic conditions below upto pH 1.5 with the action of acid they solubilize large amount of Zn, Cu, Fe, Co etc. Whatever, micro- nutrient present in the soil in elementary or complex form. The liquid inoculums of above bacteria are very effective in all types of soils. During multiplication, bacteria produce large amount of acid, hence, the pH become low up to 2.5-4.5. Use of 100ml/acres in field along with 500kg of FYM/cow dung or 100ml inoculum with 100ml its water spray on field or through drip irrigation improves the quality/yield of crops.

3.15 Manganese Solubilizers (*Penicillium citrinum*)

Penicillium citrinum a fungal culture present in soil is known to be solubilize manganese from the low grade manganese ores and from the soil if it is present. The fungus produces reductive compounds such as organic acids (oxalic acid, citric acid and which help in solubilizing of manganese. Temperature required for multiplication is 32 $\pm 2^0$C. During multiplication, pH became low up to 5.5.

3.16 Role of Liquid Biofertilizers in Various Crops

- Benefits the next crop too, due to its residual effect.
- Disease and pest occurrence reduced.
- Enhances soil health and fertility.
- Increased crop yield.
- Increases seed quality of the crops.
- Output returns economically profitable.
- Reduces chemical NPK fertilizers.

3.17 Limitations of Carrier-based Biofertilizers

- Bulk sterilization problem in terms of economic and facilities.
- Dosage contradictory to farmers.
- Even less contamination in packets the enzymatic activity of beneficial bacteria disturbed.

- Labour intensive and takes more time for quality control.
- Lack of identifiable character and instant visual effects on application.
- Less scope for export and no commercial value.
- More chance of contamination and unavailability of good carrier in local area.
- Poor moisture retention capacity and cell protection
- Problem of proper packing and high transport cost.
- Quality inconsistency of carriers.
- Restriction on use as a measure of conservation (wood charcoal).
- Shorter Shelf life (3 months) and sensitivity to temperature.
- Very slow adoption by farmers.

3.18 Methods Used for the Testing of Quality of Liquid Biofertilizers

3.18.1 General Tests

Prescribed marking on the Biofertilizer bottle like name of the product, crop for which intended, name and address of the manufacturer, batch number, date of manufacture, date of expiry, methods of use and handling instructions etc.

3.18.2 Qualitative Tests

a) **Serial Dilution Technique by Plate Count**: Total number of viable bacteria present in the Biofertilizer bottle could be known by this method, and it is completed in to following steps- Usually 10ml of Liquid based biofertilizer is removed from the bottle to be mixed in 90ml of sterile water and shaken thoroughly. After homogenization, 1ml is taken and added with 9ml of sterile water. Likewise, serially it is diluted under sterile conditions till 10^{-10}. Finally, usually 10^{-6} to 10^{-10} dilutions are chosen and out of each dilution 1.0ml of aliquot is inoculated in sterile solidified enrichment media specific to the test microorganism and spread uniformly. Petri plates are inverted and incubated in BOD incubator at +28^{0}C. The plates are observed periodically for the appearance of bacterial colonies.

b) Colonies appeared in the plates are observed for colony morphology, shapes like either round, round with scalloped margin, round with raised margin, wrinkled, concentric, irregular and spreading, filamentous, round with radiating margin, fill form, complex etc. The margin of appeared colony like either smooth, wavy, lobate, irregular, ciliate, branching, wooly, is observed. The elevation of the colony either like flat, raised, hilly, convex, drop like is observed.

c) One of the isolated colonies is removed and spread in a loop on the slide and gram stained.

d) The colonies observed are stained and its morphological characters like cellular shape, size, flagella, attachment of flagella etc. are studied preferably under a phase-contrast microscope and strain is identified.

e) Specific confirmation tests for each microorganism are done to check its authenticity.

3.18.3 Quantitative Tests

a) Through plate counts, the total number of colonies appeared are counted in an electronic colony counter and depending on the dilution factor, the colonies formed are enumerated.

b) From the broth, the total microbial populations could be counted directly under a microscope using a Petroff-Hauser Counter otherwise through a Haemocytometer and populations are enumerated. This test does not confirm the purity.

c) From the broth, samples are taken and its optical density is measured in a Spectrophotometer against its grown media as check and indirectly, the population is extrapolated through it standard.

d) The nitrogen fixing potentiality/phosphate solubilizing ability of the strains could be tested using standard procedures.

However, in the absence of adequate laboratory facilities to carry out sophisticated microbiological tests a simple "Grow out" test could be conducted to test its ability for the respective crop marked in the packet through different parameters of crop response and indirectly the quality of the biofertilizer packet is partially assessed.

Precautions

a) Don't mix with fungicides, insecticides, herbicides and chemical fertilizers.

b) Keep biofertilizers bottles away from direct heat and sunlight, and store it in cool and dry place.

c) Keep bottles away from fertilizer or pesticide containers should not mixed directly.

d) Sell only biofertilizers bottles which contain batch number, the name of the crop, the date of manufacture and expiry period. If the expiry period of the bottles is over, discard it. It will not be effective.

3.19 Liquid Biofertilizer Application Methodology

Following methods are to be used for using liquid biofertilizers such as seed treatment, root dipping and soil application method.

3.19.1 Seed Treatment Method

Seed treatment is a most common, effective and economic method as adopted for all types of inoculants. It is used for small quantity of seeds up to 5kg quantity only. The coating can be done in a plastic bag. For this purpose, a plastic bag having size (21"x10") or big can be used. The bag has to be filled with 2kg or more seeds. The bag has to be shut in such a way to trap the air as much as possible. The bag has to be twisted for 2 minutes or more until all the seed were uniformly wetted. The bag have to be opened, inflated the bag again and shaked gently. Stop shaking after each seed got a uniform layer of culture coating. The bag has to be opened and spread the seed under a shade for 20-30 minutes for dry. For large amount of seeds counting can be done in a bucket and inoculant can be mixed directly with the hand.

Seed treatment with *Rhizobium, Azotobacter, Azospirillum*, KMB along with PSM. Seed treatment can be done with any of two or more bacteria. There is no side (antagonistic) effect. The important things has to be kept in mind that the seeds must be coated first with *Rhizobium* or *Azotobacter* or *Azospirillum* when each seeds get a layer of above bacteria then PSM and KMB inoculant has to be treated on outer layer of the seeds. This method will provide maximum number of population of each bacterium required for better results. Mixing any of two bacteria and the treatment of seed will not provide maximum number of bacteria of individuals.

3.19.2 Root Dipping Method

This method is needed for the application of Azospirillum/KMB/PSM with the paddy transplanting/vegetable crops. The required quantity of Azospirillum/ KMB/PSM has to be mixed with 5-10 litres of water at one corner of the field and all the plant roots have to be dipped for minimum 1hr before sowing.

3.19.3 Soil Application Method

PSM has to be used as a soil application use 200ml of PSM per acre. Mix PSM and KMB with 400-600kg of Cow dung, FYM along with half bag of rock phosphate if available. The mixture of PSM and KMB, Cow dung and rock phosphate have to be kept under any tree or ceiling for overnight and maintain 50% moisture (Chandra, 2015). Use the mixture as a soil application in rows or during leveling of soil.

Table 3.4: Recommended dosage of Liquid Biofertilizer in different Crops

Field Crops	Recommended Biofertilizer	Application Method	Quantity to be Applied
Pulses: Chickpea, pea, Ground nut, Soybean, beans, Lentil, lucern, Berseem, green gram, Black gram, cowpea, Pigeon pea.	*Rhizobium*	Seed treatment	200ml/acre
Cereals: Wheat, oat, barley	*Azotobacter/ Azospirillum*	Seed treatment	200ml/acre
Rice	*Azospirillum*	Seedling treatment	200ml/acre
Oil Seeds: Mustard, Sesame Linseeds, Sunflower, Castor	*Azotobacter*	Seed treatment	200ml/acre
Millets Pearl millets, Finger	*Azotobacter*	Seed treatment	200ml/acre
Maize & Sorghum	*Azospirillum*	Seed treatment	200ml/acre
Forage Crops & Grasses: Bermuda grass, Sudan grass, Napier Grass, Para grass and Star grass etc.	*Azotobacter*	Seed treatment	200ml/acre
Other Misc. Plantation Crops: Tobacco	*Azotobacter*	Seedling treatment	500ml/acre
Tea, Coffee	*Azotobacter*	Soil treatment	400ml/acre
Rubber, Coconuts	*Azotobacter*	Soil treatment	2-3ml/acre
Agro-Forestry/Fruit Plants: All fruit/agro-forestry (herb, shrubs, annuals & perennial) plants for fuel wood, fodder, fruits, gum, spice, leaves, flowers, nuts & seeds purpose	*Azotobacter Azospirillum*	Soil treatment	Soil 1-2 ml/plant at Nursery Soil 2-3 ml/plant at nursery
Leguminous plants / Trees	*Rhizobium*	Soil treatment	Soil 1-2 ml/plant

Source: Panwar and Jain, 2015

Note: Dosage recommended when count of Inoculum is 1×10^9cells/ml. If the population is less than 10^8 cells/ml dosage will be ten times more besides above said Nitrogen fixers, Phosphate solubilizers and Potash mobilizers at the rate of 200ml/acre could be applied for all crops.

3.20 Crisis of Efficient Strains

Unavailability of potential regional strains is one of the major reasons. The specificity and competitive ability of the strain is the key point on which the efficacies of the organism relay with respect to the hosting soil and plant variety. The ability to fix nitrogen and survival capacity are the limiting factors of the liquid as well as the carrier based biofertilizers.

3.21 Possible Genotypic Changes

During the production of the fertilizers the organism may get interacted with other organisms which may leads to change in basic character of the organisms. Apart from this during fermentation the strains may undergo mutations which may alter the efficacy and viability leading to the economic loss.

3.22 Lack of Awareness

In spite of many ongoing projects on the development of biofertilizers, proper attention towards the technology is still needed in order to manifest the results at field level. Communication gap and miscommunication of the farmers by some commercial producers also is a considerable constraint. Lack of storage and marketing facilities, consistent awareness in the farmers by conducting community programs is still welcome (Revellin *et al.*, 2001). As the production and distribution of biofertilizers is seasonal, the commercial production units may suffer from lack of demand. Moreover, the establishment of microorganisms resisting the antagonistic activity of the native microbes is always challenging. Soil nature, temperature, humidity, pH, local insect population, etc., are some other acclimatization problems (Ngampimol and Kunathigan, 2008).

The formulated liquid biofertilizers of Rhizobium, Azotobacter, Azospirillum and PSB (*Bacillus megaterium*) were stored in BOD incubator at 28±2 °C for a period of 180 days and colony forming units were determined at monthly intervals. The liquid biofertilizers formulated using PVP in addition to glycerol at the rate of 0.5% each retained maximum number of colonies in all strains followed by PEG (Santosh, 2015),

3.23 Future Strategies

- Constructive awareness and technical support by microbiologists and agricultural professionals must be provided to the Agrarians.
- Development of new strains with enhanced capabilities by genetic engineering techniques and rDNA technology is needed to maintain an eco-friendly and sustainable agriculture.
- Exploring the novel soil bacteria and maintaining the genomic libraries for future exploitation.
- Genotypic study of the strains and molecular characterization of the plant parts is necessary to understand the plant mechanisms.
- Global standards in the research & development should be maintained during the production and storage of the formulations.

- Identification and characterization of potential organisms (unexplored) and their effective exploitation in the field of Agriculture is urgently needed.
- Selection and application of suitable bio-inoculants with respect to soil nature, agro climatic condition, crop variety under proper agriculture practices is needed.
- Soil analysis, crop rotation, organic manure usage, maintenance of proper moisture content, regular sterilization practices are emphasized which are necessary to maximize the biofertilizer efficacy.
- Study of soil texture and compatible studies with respect to microbial interventions.
- Suitable combinations of microbial formulations (liquid microbial consortium) with optimized field results are preferable for the sustainable production.

References

Ahemad, M. and Khan, M.S. (2011). Pest Manag. Sci 67: 423-429

Anubhav Liquid Biofertilizer (Source: http://www.icar.org.in/en/node/5667 Technology for Liquid Bio-Fertilizers Commercialized)

Bot, A.; Benites, J. (2005). The importance of soil organic matter. Natural resources management and environment department, FAO corporation document repository. http://agricoop.nic.in/Annual%20report2011-12/ARE.pdf.

Chandra, K., Singh., T.; Srivathsa., H. (1999). In: Biofertilizer ready recokner Pages. Published by – Regional Biofertilizer Dev. Centre, Bhubaneswar, Orissa.

Chandra, K.; Greep, S.; Srivathsa, R.S.H. (2005). Promotion of organic farming in Southern States – role of regional centre of organic farming: In proceeding of the Seminar on organic agriculture in peninsular India – promotion, certification and marketing. pp 25-27.

Chandra, K.; Greep, S.; Srivathsa, R.S.H. (2005). Spices India 18(7): 29-33.

Chandra, K. (2015). NPK-Liquid Biofertilizers (Poly Culture) Biofertiliser Newsletter, National Centre of Organic Farming Ghaziabad U.P.

Eckert, D. (2012). Efficient Fertilizer Use- Nitrogen. Retriewed from http://www.rainbowplant food.com/agromomics/efu/nitrogen.pdf.

Gautam, P.; Dashora, L.N.; Solanki, N.S.; Meena R.H.; Upadhyay B. (2017). Int. J. Curr. Microbiol. App. Sci 6(12): 922-927.

Ghosh N; (2014). Ecological Economics 49: 149-162.

Khan, M.S.; Zaidi, A.; Wani, P.A. (2009). Biomedical and Life Sciences 551-570.

Muthusamy, K,; Gopalakrishnan, S.; Ravi, T.K.; Sivachidambaram, P. (2008). Current Science, 94: 736- 774.

Ngampimol, H.; Kunathigan, V. (2008). Aus. J. T. 11: 204-208.

Panwar, J.D.S.; Jain, A.K. (2015). Organic Farming- Scope and Use of Biofertilizers, New India Publishing Agency New Delhi.

Patel, B.C. (2011). Advance method of preparation of bacterial formulation using potash mobilizing bacteria that mobilize potash and make it available to crop plant. WIPO Patent Application WO/2011/154961.

Pindi, P.K.; Satyanarayana, S.D.V. (2012). J. Bioferil. Biopestici. 3(4): 1-9.

Revellin, C.; Giraud, J.J.; Silva, N.; Wadoux, P.; Catroux, G. (2001). Pest Manag. Sci. 57: 1075-1080.

Santhosh, G.P. (2015). International Journal of Researches in Biosciences, Agriculture and Technology 2(7): 243-247.

Zaidia, A.; Khan, M.S.; Aamila, M. (2004). J. Plant Nutr 27: 601-612.

4

Earthworm and Vermitechnology

Earthworms have been on the earth for over 20 million years. In this time they have faithfully done their part to keep the cycle of life continuously moving. Their purpose is simple but very important. They are nature's way of recycling organic nutrients from dead tissues back to living organisms. Many have recognized the value of these worms. Ancient civilizations, including Greece and Egypt valued the role of earthworms played in soil. The Egyptian Pharaoh, Cleopatra said, "Earthworms are sacred"

4.1 Earthworms

Earthworms belong to Phylum Annelida of animal Kingdom. They are long and cylindrical shape, length of earthworms vary from a few mm to nearly 3mtr as noted in case of *Megas colex* species found in Australia. The body of earthworms has a large number of grooves. Due to these grooves the body of the earthworm is divided into number of compartments called segments. In mature earthworms, few anterior segments appear swollen due to glandular thickening. This part is called as clitellum. The segments are not clearly seen in this part. There are 2500 to 3,000 species of earthworms in the world which are adapted to range of environment. More than 350 species have been identified in India. The species of earthworms are differentiated mostly on the basis of external features such as length of earthworms, number of segments, position of clitellum and external aperture present on the body. Earthworms were referred by Aristotle as "the intestines of earth and the restoring agents of soil fertility" (Shipley, 1970). They are biological indicators of soil quality (Ismail, 2005), as a good population of earthworms indicates the presence of a large population of bacteria, viruses, fungi, insects, spiders and other organisms and thus a healthy soil (Lachnicht and Hendrix, 2001). The role of earthworms in the recycling of nutrients, soil structure, soil productivity and agriculture, and their application in environment and organic waste management is well understood (Edwards *et al.*, 1995; Tomlin *et al.*, 1995; Shuster *et al.*, 2000; Ansari and Ismail, 2001a,b; Ismail, 2005; Ansari and Ismail, 2008; Ansari and Sukhraj, 2010).

4.2 Life cycle of Earthworms

Earthworms are hermaphrodite. However, single earthworm cannot usually self-fertilize itself and two mature earthworms are required to propagate. At the time of egg lying, the clitelium is transformed into hard, girdle like capsule called cocoon. Shedding of cocoon takes place about 10 days after fertilization. Cocoons are small, spindle shaped and translucent structures. The size of cocoon varies with the species of earthworms. The fertilization takes places inside the cocoon. Although, number of fertilized ova in each cocoon ranges from 1 to 20, only a few of them survive and hatch. The juvenile earthworms crawl out of cocoons. The cocoons have an incubation period of 18 days after which they hatch into earthworm. The entire life cycle from hatching of cocoons, maturing of juveniles and again formation of cocoons take a period of 50 to 60 days. Normally the average life span of an earthworm is about one year during which they reproduce about 10 times.

4.3 Type of Earthworms

Ecologically, earthworms are classified into following three types (Table 4.1) such as **(a)** Epigeic **(b)** Anecic, and **(c)** Endogeic (Card *et al.,* 2004).

4.3.1 Epigeics (Greek for "upon the earth")

These "decomposers" are the type of worm used in vermicomposting. These are surface feeders and feed on leaf litter and other plant wastes. They do not have permanent burrows. They do not mix the soil and generally have no effect on soil structure etc. They have affinity for nitrogen rich organic matter, and live in more unstable environment. They resemble the stages of ecological development, characterized by small body size, high metabolic rate, high fecundity and short life cycle. These earthworms are normally from forest floor community in tropical countries. With loss of natural forests, they have moved into plantations. When such earthworms are chosen to work on waste organic matters their major contribution is in reducing the particle size of organic matter.

4.3.2 Anecic (Greek for "out of the earth")

These are burrowing worms that come to the surface at night to drag food down into their permanent burrows deep within the mineral layers of the soil. They feed on leaf litter and soil of upper layers. They move along vertically into the soil forming vertical furrows. Vermicasts are liberally produced on the soil surface.

4.3.3 Endogeic (Greek for "within the earth")

These are also burrowing worms but their burrows are typically shallower and they feed on the organic matter already in the soil, so they come to the surface only rarely. They live within the soil and ingest large quantities of organically rich soil. The generally make horizontal furrows and deposit their casts within the furrow.

Epigeic and anecic type of earthworms are important in management of organic wastes e.g. compost formation. Epigeic such as *Eisenia foetida* and *Eudrilus euginiae* are exotic and *Perionyx excavatus* is native which are being used for vermicomposting. There are three varieties of earthworms available in India which are called manure worms. *Eisenia foetida* and *Eudrilus euginiae* are exotic species and used in composting. *Perionyx excavatus* and *Lampito mauritti,* epigeic and anecic forms respectively have also been successfully used in vermiculture and vermicomposting. Anecic such as *Lampito mauritti* is a native one and can play a role *in situ* biodegradation of organic wastes of soil. Endogeic commonly include *Metaphire posthuma* and *Octochaetonia thurstoni* which are indigenous and used for vermicomposting. Research work have been done by Kaplan *et al.* (1980) to show that it grows in a wide range of agricultural wastes. *Eisenia fetida* worms have also been shown to have a wide tolerance range for temperature from 0° to 40°C (Phillip, 1981) and suitability for vermicomposting in terms of temperature (Reinecke *et al.* 1992) and moisture (Reinecke and Venter, 1987).

Table 4.1: Comparative Chart of the Major Characteristics of three Earthworm Genera

Sl. No.	Characteristics	A*	B'	C**
		Perionyx excavatus (Hallatt et al. (1990)	Eudrilus eugeniae (Viljoen and Reinecke, 1989)	Eisenia foetida (Venter and Reinecke, 1988)
1.	Growth rate (mg/worm/day)	3.5	12.0	7.00
2.	Maximum body mass (mg/worm)	600.0	4294.0	1500.00
3.	Maturation attained at age (days)	21.0	40.0	150.00
4.	Start of cocoon production (days)	24.0	46.0	55.00
5.	Cocoon production (worm/day)	1.1	1.3	0.35
6.	Cocoon incubation period (days)	18.7	16.6	23.00
7.	Mean number of hatchings	1.1	2.7	2.70
8.	Number of hatchings (cocoon^{-1})	1.3	1.5	1.90
9.	Duration of life-cycle (days)	46.0	60.0	70.00

Source: *Gaur (2010) **Viljoen and Reinecke (1989)

4.4 Role of Earthworms in Nature

4.4.1 Soil Hydraulic Properties

By burrowing through soil, earthworms alter the soil fabric which, in turn, changes the physical characteristics of soil. *Allolobophora caliginosa* and *Allolobophora rosea* give burrow formation on soil hydraulic properties and shows saturated hydraulic conductivity as well as percolation rates was higher when compared to the control.

4.4.2 Soil Amelioration

Earthworms play an important role in the process of soil amelioration and maintenance of soil fertility. They incorporate organic matter and turn over large amount of soil by burrowing, feeding and casting. It leads to improved soils structure (Strewart and Scullion. 1988). Earthworm activities are advantageous for the stabilization of reformed soil aggregates. The organic carbon content was increased by 4.1 to 21.0% for burrow wall material and by 21.2 to 43.0% for casts. The total content of polysaccharides was increased by 35 to 87% for casts and by 33 to 46% for the burrow-wall material of all earthworm species (Zhang and Schrader, 1993).

4.4.3 Soil Formation

Darwin (1981) first recognized that earthworms are common in the A-l horizon of soil, and was called as vegetable mould. These result from the burrowing, casting and subsequent mixing of soil mineral and organic materials by earthworms. In most soils, earthworms burrow and cast throughout the A and B horizons and are sometimes below the B horizon. Earthworms are usually in high numbers near the soil surface while the numbers are relatively low in deeper soil layers

4.4.4 Organic Carbon Turnover

Allobophora caliginosa alters the biomass and the metabolic activity of the edaphic microflora over a wide range of soil, and these alterations are highly soil and resource specific and the microbial carbon turnover in soils by *A. caliginosa* is significantly different from the microbial carbon turnover in aged casts and burrows (Wolters and Joergensen, 1992). Vermicomposting resulted in significant reduction in C/N ratio and increase in mineral nitrogen after 90 days of composting over treatments uninoculated with earthworms (Bansal and Kapoor, 2000)

4.4.5 Recycling of Organic Matter

The role of worms in organic matter recycling process and their effect on acceleration of decomposition process (Heungens, 1969) is gaining importance.

4.4.6 Availability of N and P to Plants

Earthworms have an important role in transfer of nitrogen from decaying plant tissues by releasing it in chemical form that can be easily taken up and recycled for plant growth. The significance of earthworms in the nitrogen economy of soils, especially the role of excreted nitrogen contained in mucus and that derived from the decomposition of dead earthworm tissues has received considerable attention. Phosphorus is limiting element for plant growth and especially for fruit and seed production. Much of the phosphorus in soils is bound to organic matter in forms that are unavailable to plants. It is now well established that passage through the gut of some lumbricid earthworms results in some of this phosphorus being converted to forms that are available to plants. This process involves uptake of phosphorus by bacteria and fungi from a pool of phosphorus that is fixed by soil minerals and subsequent ingestion of these organisms by soil fauna like earthworms. Some release of P in forms available to plants is then mediated by the enzyme phosphatase(s) that are produced within the earthworm gut, while further release of P may be induced by micro-organisms in casts after their excretion (Lee, 1992). Vermicomposting may prove to be an efficient technology for providing better P nutrition from different organic wastes (Gosh *et al.,* 1999).

4.4.7 Exchangeable K

Casts of earthworm contained two to three times more available K than the surrounding soil (Bezbordov and Khalbayeva, 1990; Hullegalle and Ezumah, 1991). Basker *et al.,* (1993) reported that the changes in the K dynamics of soil resulting from ingestion by worms must be due to atleast in part to the changes in the distribution of K between exchangeable and non-exchangeable forms.

4.4.8 Plant Nutrient Availability

Earthworms increased significantly in number and the rate of growth and yield of plants growing on the inoculated soils. Edwards and Lofty (1980) demonstrated that increased root growth after earthworm inoculation was primarily due to the formation of burrows that were responsible for more available nutrients than the surrounding soil. They have clearly shown that burrows provided ideal channels which promoted the nutrient availability for buffer as well as maximum root growth. Better root growth is also an index for efficient utilization of available nutrients which in turn reflect on the biomass productivity of crops.

4.4.9 P Solubilization

Soil micro-organisms particularly bacteria and fungi associated in the root regions of plant play an important role in supply of phosphorus (Gaur and Mathur, 1990). Earthworm casts are reported to improve the soil fertility (Krishnamoorthy and Vajranabhaiah, 1986). The soil fertility factor in these casts might be due to microbial enzymes which help in the availability of nutrients. Phosphate solubilization by fungi isolated from earthworm gut was studied by employing dicalcium and tricalcium phosphates. *Choanephora cucurbitarum* and *Verticillium chlamydosporum, Neurospora crassa* were responsible for liberation of maximum phosphorus form dicalcium and tricalcium phosphates. Fungi differ in their capacity to solubilize phosphorus form dicalcium and tricalcium phosphates. *Verticillium chlamydosporum* was most efficient and dissolved maximum amount of phosphate in soil and made available to plants (Dorcas *loy et al.* 1991).

4.4.10 Microbial Enzymatic Activities

The interaction between earthworms and microorganisms are of major importance in the degradation of organic matter and release of mineral nutrients into soil (Lee, 1985). Certain micro-organisms are killed during passage through gut (Yeates, 1981) whereas other proliferates. Efficient dispersal of beneficial organisms like *Rhizobium, Pseudomonas* (Madsen and Alexander, 1982), *Frankia* spp. (Reddell and Spain, 1991) and hyphal filament spores of AM fungi (Rebatin and Stinner, 1988) and possibility of biocontrol agent application through earthworms food were reported by Duble *et al.,* (1994). Earthworms have also been reported to increase the availability of plant nutrient in soil (Krishnamoorthy and Vajranabhiah, 1986) increase the production of plant growth regulators (Tomati *et al.,* 1988) and also increase foliar concentration of Ca, Na, Mn, Cu, Fe and Al in wheat (Stephens *et al.* 1994).

4.5 Vermicomposting

Composting is one of the feasible means for converting bio-degradable solid wastes into beneficial organic soil amendments for supporting environment friendly agricultural production system. Many beneficial organisms and microorganisms act as chemical decomposer in the process of formation of stable organic end-products (compost) during composting. Among them, decomposers like earthworms play significant role in stimulating the process of composting, enhancing nutrient value while fastening the process of stable organic end-product formation. This process of involvement of earthworms in preparing enriched compost is called vermicomposting. It is one of the simplest methods to recycle agricultural wastes and to produce quality compost. Earthworm acts physically an aerator, crusher and mixer, chemically a degrader and biologically a stimulator in the process of decomposition. Earthworms consume biomass (decaying organic matter) and excrete it in a digested form called as worm casts or worm manure. Worm casts are popularly called as black gold. They are rich in essential plant nutrients, plant growth promoting substances, beneficial soil micro flora and having properties of inhibiting pathogenic microbes. As a result, the organic endproducts produced by the use of earthworms i.e. vermicompost also inherits most of the beneficial properties (to soil health and crop productivity) of black gold. Vermicompost acts as an organic soil amendment- improves three dimensional soil health's (physical, chemical & biological properties). On application of vermicompost, it enhances the soil quality by improving its physicochemical and biological properties. The earthworm's underground burrows modify soil hydro-thermal and aeration regimes by making the soil more porous thus, allowing free movement of air, infiltration of water into deeper soil layers for better profile moisture recharge and root water uptake processes (Chanu *et al.,* 2018). Vermicompost is becoming popular as one the major components of the organic farming system because of its high nutritive value in addition to an important organic soil amendment. Vermitechnology can be used as a potential weapon. Earthworms have the capacity to stabilize a wide range of toxic organic wastes into value added products. When vermicompost is applied in the field, it enhances the quality of soil through increasing microbial biomass and microbial activity which play a major role in nutrient cycling (Boruah, 2019), in soil waste management (Iyyanki *et al.,* 2017) and earthworms converts domestic and food industry wastes into biofertilizer (Rorat and Vandenbulcke, 2019).

4.6 Terms related to Vermicomposting

4.6.1 Vermiculture: Vermiculture is the scientific method of breeding and raising earthworms under controlled conditions (mostly hydrothermal regimes).

4.6.2 Vermitechnology: Vermitechnology means the combination of vermiculture and vermicomposting.

4.7 Materials for Preparation of Vermicompost

Several types of biodegradable wastes as crop residues, weed biomass, vegetable waste, leaf litter, hotel refuse, waste from agro-industries and biodegradable portion of urban and rural wastes are used for the preparation of vermicompost.

Preparation of vermicompost - at a glance

Vermicompost is obtained by turning organic debris and residues to compost using earthworms. Earthworms can feed on many types of organic waste like agricultural waste, forest litters, kitchen waste, etc. The organic wastes after entering the earthworms' alimentary canal undergo some chemical changes rendering it odorless and neutral. Vermicompost production unit can be set up in any land which is not under any economic use but shady and free from water stagnation. The site should also be nearer to a water resource. Vermicomposting production unit may be set up in cement or brick tank, wooden boxes, plastic bins, silpaulin bags of varying dimensions (preferable size of 10′x4′x2′), a basket, a bucket and even in soil pits. A quality vermicompost can be produced by using the suitable earthworm (350 to 360 worms per m3 of bed volume); bedding material that will provide the worms a relatively stable habitat and feeding with good worm food. Utmost care should be taken to maintain adequate moisture (about 60% water content by weight), aeration, and temperature (28 – 35^0C) during the composting period.

4.8 Requirements for Vermicomposting

4.8.1 Basic Raw Material

Any types of biodegradable wastes like crop residues, weed biomass, vegetable waste, leaf litter, hotel refuse, waste from agro-industries and biodegradable portion of urban and rural wastes can be used as basic raw materials for vermicomposting. A mixture of leguminous and non-leguminous crop residues enriches the quality of vermicompost.

4.8.2 Selection of Suitable Earthworm

Only surface dwelling earthworms should be used for vermicomposting.

4.8.3 Starter: Cow Dung, Biogas Slurry, or Urine of Cattle

During the beginning of the process of composting, cow dung can be used as feeding material in order to breed sufficient numbers of earthworms. On attaining desired number of worm population, subsequently other sources of organic wastes can be provided to maintain the population of earthworms.

4.8.4 Site Selection

Vermicompost production can be done in any place which is having shades, high humidity and cool. Abandoned cattle shed, or poultry shed or unused buildings can also be used. If it is to be produced in the open area, artificial shading should be provided. The waste heaped for vermicompost production should be covered with moist gunny bags.

4.8.5 Containers for Vermicompost Production

A cement tub may be constructed to a height of 2½ feet and a breadth of 3 feet. The length may be fixed to any level depending upon the size of the room. The bottom of the tub is made sloppy to drain the excess water from vermicompost unit. A small sump is necessary to collect the drain water. If hard floor is used, hollow blocks/bricks may be arranged in a compartment to a height of one foot, breadth of 3 feet and length to a desired level to have quick harvest. In this method, moisture assessment will be very easy. Vermicompost can also be prepared in wooden boxes, plastic buckets, silpaulin bag or in any containers with a hole at the bottom for draining excess water.

4.9 Role of Earthworm in Vermicomposting

For vermicompost production, the surface dwelling earthworm alone should be used. The earthworm, which lives below the soil, is not suitable for vermicompost production. The African earthworm (*Eudrillus engenial),* Red worms (*Eisenia foetida)* and composting worm (*Peronyx excavatus)* are promising worms used for vermicompost production (Fig. 4.1 A, B & C). All the three worms can be mixed together for vermicompost production. The African worm (*Eudrillus engenial)* is preferred over other two types, because it produces higher production of vermicompost in short period of time and younger ones in the composting period.

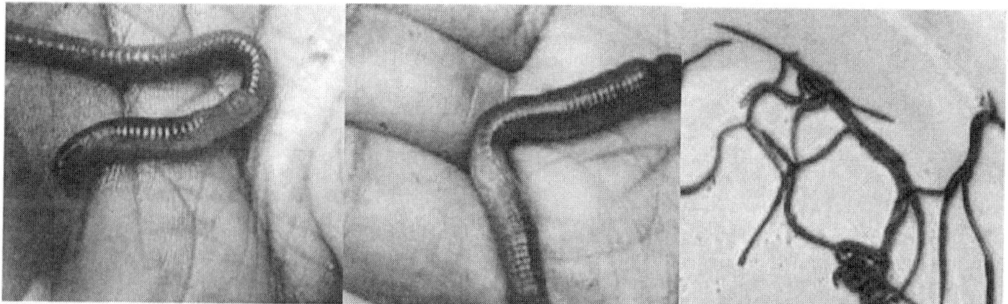

Fig 4.1A. African Earthworm (*Eudrillus euginiae*)

Fig 4.1B. Tiger Worm or Red Wrinkle (*Eisenia foetida*)

Fig 4.1C: Asian Worms (*Perionyx ecavatus*)

4.9.1 Favorable conditions for earth worm culture in the composting material

- pH: Near neutral (range between 6.5 to 7.5).
- Moisture: 60-70% of the moisture (wt./wt.); below and above this range, mortality of worms taking place.
- Aeration: 50% aeration from the total pore space.
- Temperature: Range from 18 to 35^{0}C.

4.10 Methods of Vermicomposting

Vermicomposting is done by various methods. Among them, bed and pit methods are more common.

4.10.1 Bed Method: Composting is done on the pucca/kachcha floor by making a bed (dimension: 6 x 2 x 2 feet) of organic mixture. This method is easy to maintain and to practice, and step wise procedure is given here

a) Processing involves collection of wastes, shredding, mechanical separation of the metal, glass and ceramics and storage of organic wastes.

b) Pre-digestion of organic waste for twenty days by heaping or dumping the material along with cattle dung slurry. This process partially digests the material and fit for earthworm consumption.

c) Preparation of earthworm bed. A concrete base is required to put the waste for vermicompost preparation. Loose soil will allow the worms to go into the soil and also while watering; all the dissolvable nutrients go into the soil along with water.

d) A layer of 15-20 cm of chopped dried leaves/grasses should be kept as bedding material at the bottom of the bed.

e) Beds of partially decomposed material of size 6x2x2 feet should be made. Each bed should contain 1.5-2.0 q of raw material and the number of beds can be increased as per raw material availability and requirement.

f) Red earthworm (350 -360 worms per m^3 of bed volume) should be released in the upper layer of the bed.

g) Water should be sprinkled with can immediately after the release of worms.

h) Beds should be kept moist by sprinkling of water (daily) and by covering with gunny bags/polythene.

i) Bed should be turned once after 30 days for maintaining aeration and for proper decomposition.

j) Compost gets ready in 45-50 days.

k) The weight of the finished product is about 75% of the raw materials used.

4.10.2 Pit Method: Composting is done in the cemented pits, wooden boxes, plastic buckets, silpaulin bag, baskets, etc. The unit is covered with thatch grass or any other locally available materials. Step wise procedure is given here

a) Pit size of dimensions 10′x 4′ x 2′ of either cement or vermibag is maintained. The length and width can be increased or decreased depending upon the availability of material but not the depth because the earthworms' activity is confined to 2 feet depth only.

- 1st layer: bedding material of 1" thick with soft leaves
- 2nd layer: 9" thick organic residue layer finely chaffed material
- 3rd layer: dried cattle dung + water equal mixture of 2" layer.

b) The layer is continued until the pile is filled up.

c) On 25 days old unit, 795-820 worms are introduced into the pit (350 -360 worms per m3 of bed volume) without disturbing the pit.

d) Proper moisture and temperature is maintained by frequent watering, turnings and subsequent staking.

e) The turnover of the compost is 75% (If the total material accommodated in the pit is 1000 kg; the out turn will be 750 kg).

f) The filled materials are watered and turned at regular interval.

4.10.3 Recomposting and *In-situ* Vermicomposting

Recomposting is done in the same pit or bed following the same steps as described in the above mentioned pit/bed methods. In-situ vermicomposting can be done by direct field application of vermicompost at 5 t/ha followed by application of cow dung (2.5 cm thick layer) and then a layer of available farm waste about 15 cm thick. Watering should be done at an interval of 15 days.

4.11 Handling and Harvesting of Vermicompost

Handling and harvesting of earthworm and vermicompost is an easy task, but it needs little attention, otherwise it may disturb the earthworm activity and also may delay the process of vermicomposting. Vermicomposting is a 5 phase process, needs care at each and every stage to get a quality vermicompost for fetching remuneration on a sustainable basis. The commonly faced problems, management options, different phases of vermicomposting, and precaution to be taken care during different phases of composting are listed here..

4.11.1 Commonly Faced Problem in Vermicomposting

Vermicomposting is more sensitive than other composting methods and may induce to the following problems:

- **Extreme weather condition**: Vermicompost is susceptible to extreme weather conditions such as frost, heavy rainfall, drought and overheating.

- **Putrefication:** Anaerobic conditions (due to compaction and lack of oxygen) can quickly lead to putrefication.
- **Predators:** ants, birds, lizards may disturb the activity of earthworm.

4.11.2 Five Phases of Vermicomposting

Phase 1: Initial Process: Processing involving collection of wastes, shredding, mechanical separation of the metal, glass and ceramics and storage of organic wastes.

Phase 2: Pre-digestion of organic waste: Pre digestion of organic waste for twenty days by heaping the material along with cattle dung slurry. This process partially digests the material and fit for earthworm consumption. Cattle dung and biogas slurry may be used after drying. Wet dung should not be used for vermicompost production.

Phase 3: Preparation of earthworm bed: A concrete base is required to put the waste for vermicompost preparation. Loose soil will allow the worms to go into soil and also while watering; all the dissolvable nutrients go into the soil along with water.

Phase 4: Collection of earthworm after vermicompost collection: Sieving the composted material to separate fully composted material. The partially composted material will be again put into vermicompost bed.

Phase 5: Storing the vermicompost: Storing the vermicompost in proper place to maintain moisture and allow the beneficial microorganisms to grow.

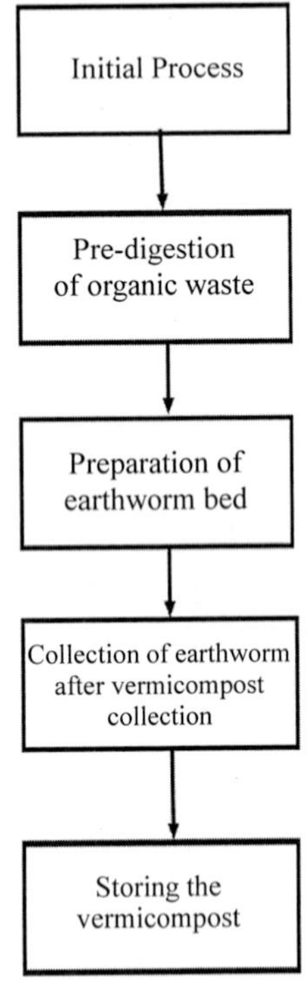

Fig. 4.2: Phases of Vermicomposting

4.11.3 Precautions to be Taken During the Five Phases

4.11.3.1 Collection of waste material: The collected waste material should be processed for shredding, mechanical separation of the metal, glass and ceramics and should be stored in a proper place.

4.11.3.2 Pre-digestion: Pre-digestion of organic waste should be done for at least 20-25days by heaping the material along with cattle dung slurry and regular watering. This process partially digests the material and fit for earthworm consumption. Addition of higher quantities of acid-rich substances such as citrus wastes should be avoided. Any organic wastes – cow dung, crop residues, farm wastes, vegetable market wastes, and fruit wastes can be used as a raw material for composting. Use of wet dung should be avoided for vermicompost production. At least 20-25 days old cow dung should be used to avoid excess heat generation.

4.11.3.3 Prepration of Earthworm bed and composting: The earthworm bed propared for vermicomposting must ensure to obtain the qualitative vermicompost from short span of time for this, all basic neccerities are to be used

4.11.4 Essentials requirements for Compost Worms-Need Majority Five Basic Things

The earthworm bed prepared for vermicomposting must ensure the five basic things to obtain quality vermicompost from a short span of time.

a An hospitable living environment, usually called "bedding";

b A food source;

c Adequate moisture (greater than 50% water content by weight);

d Adequate aeration;

e Protection from temperature extremes.

These five essentials are discussed in more detail below.

4.11.4.1 Bedding

Bedding is any material that provides the worms a relatively stable habitat. This habitat must have the following characteristics:

a) High absorbency

Earthworms breathe through their skins and therefore, must have a moist environment of living. Worm dies if their skin dries out. The bedding material must be able to absorb and retain water fairly enough if the worms are to be thrived.

b) Good bulking potential

The flow of air is reduced or eliminated if the material is too dense to begin with, or packs too tightly. There should be proper aeration as worms require oxygen to live, just as we do. A variety of factors, including the range of particle size and shape, texture, strength and rigidity of the materials affect the overall porosity of the bedding.

c) Low protein and/or nitrogen content (High Carbon: Nitrogen ratio)

Bedding material with high Carbon: Nitrogen ratio is desirable as high protein/ nitrogen levels can result in rapid degradation, heating creates an inhospitable environment for the worm. Heating can occur safely in the food layers of the vermicomposting system, but not in the bedding.

4.11.4.2 Vermiculture Bed

Vermiculture bed or worm bed (3 cm) can be prepared by placing saw dust, straw, coir waste, sugarcane trash etc. at the bottom of tub/container. A layer of fine sand of 3 cm thick should be spread over the culture bed followed by a layer of garden soil (3cm). All layers must be moistened with water. In case of bed method, the floor of the unit should be compacted to prevent earthworms' migration into the soil.

Table 4.2: Common Bedding Materials

Bedding Material	Bulking Pot.	Absorbency	C:N Ratio
Hay – general	Medium	Poor	15 - 32
Straw – general	Medium-Good	Poor	48 - 150
Straw – oat	Medium	Poor	48 - 98
Straw – wheat	Medium-Good	Poor	100 - 150
Corn stalks	Good	Poor	60 - 73
Corn cobs	Good	Poor-Medium	56 - 123
Corn Silage	Medium	Medium-Good	38 - 43
Bark – hardwoods	Good	Poor	116 - 436
Bark -- softwoods	Good	Poor	131 - 1285
Sawdust	Poor-Medium	Poor-Medium	142 - 750
Shrub trimmings	Good	Poor	53
Hardwood chips, shavings	Good	Poor	451 - 819
Corrugated cardboard	Medium	Good	563
Lumber mill waste -- chipped	Good	Poor	170
Paper fibre sludge	Medium	Medium-Good	250
Paper mill sludge	Medium	Good	54
Newspaper	Medium	Good	170
Paper from municipal waste stream	Medium	Medium-Good	127 - 178
Softwood chips, shavings	Good	Poor	212 - 1313
Leaves (dry, loose)	Poor-Medium	Poor-Medium	40 - 80
Horse Manure	Good	Medium-Good	22 - 56
Peat Moss	Medium	Good	58

4.11.4.3 Worm Food Source

Under ideal conditions earthworms are able to consume in excess of their body weight each day, although in general they consume ½ of their body weight per day. They feed on anything organic that is, of plant or animal origin, manures are the most commonly used worm feedstock. Dairy and beef manures are generally considered as the best natural food for *Eisenia foetida*, with the possible exception of rabbit manure. The former, being more often available in large quantities, is the feed most often used.

Table 4.3: Common Worm Feed Stocks

Sr. No.	Food	Advantages	Disadvantages
1	Legume hays	Higher N content makes these good feed as well as reasonable bedding.	Moisture levels not as high as other feeds, requires more input and monitoring
2	Pre-composted food wastes	Good nutrition	Nutrition less than with fresh food wastes.
3	Fresh food scraps (e.g., peels, other food prep waste, leftovers, commercial food processing wastes)	Excellent nutrition having good moisture content, and possibility of revenues from waste tipping fees	Pre-composting is required
4	Seaweed	Good nutrition due to high in micronutrients and beneficial microbes	Salt must be rinsed off
5	Biosolids (human waste)	Excellent nutrition and possibility of waste management revenues	Heavy metal and/or chemical contamination
6	Cattle manure	Good nutrition	Weed seeds make pre-composting necessary
7	Sheep/Goat manure	Good nutrition	Require pre-composting (weed seeds) and small particle size can lead to packing
8	Poultry manure	High N content results in good nutrition and a high-value product	High protein levels can be dangerous to worms,
9	Rabbit manure	N content second only to poultry manure therefore contains very good mix of vitamins & minerals	Must be leached prior to use because of high urine content.
10	Hog manure	Good nutrition and produces excellent vermicompost	Usually in liquid form, therefore must be dewatered or used with large quantities of highly absorbent bedding

Sr. No.	Food	Advantages	Disadvantages
11	Fish, poultry, blood wastes, animal mortalities	High N content provides good nutrition; opportunity to turn problematic wastes into high-quality product	Must be pre-composted until past thermophillic stage
12	Corrugated cardboard (including waxed)	Excellent nutrition due to high-protein glue used to hold layers together	Must be shredded (waxed variety) and/or soaked (non-waxed) prior to feeding

- Horse dung, due to the risk of Tetanus virus, lethal to human beings is not advisable to be used as feeding material for earthworms.
- Paddy husk, marigold and pine needles should not be used as feeding materials for earthworms.

4.11.4.4 Moisture and Aeration

The bedding used must be able to hold sufficient moisture if the worms are to have a livable environment. They breathe through their skins and moisture content in the bedding of less than 50% is dangerous. With the exception of extreme heat or cold, nothing will kill worms faster than a lack of adequate moisture. The ideal moisture-content range for materials in conventional composting systems is 45-60%. In contrast, the ideal moisture-content range for vermicomposting or vermiculture processes is 70-90%.

Worms are oxygen breathers and cannot survive anaerobic conditions (absence of oxygen). When factors such as high levels of grease in the feedstock or excessive moisture combined with poor aeration conspire to cut off oxygen supplies, areas of the worm bed, or even the entire system, can become anaerobic. This will kill the worms very quickly. Not only are the worms deprived of oxygen, they are also killed by toxic substances (ammonia) created by different sets of microbes that bloom under these conditions. This is one of the main reasons for not including meat or other greasy wastes in worm feedstock unless they have been pre-composted to break down the oils and fats.

4.11.4.5 Temperature Control

Controlling temperature to within the worms' tolerance is vital to both vermicomposting and vermiculture processes. This does not mean, however, that heated buildings or cooling systems are required. Worms can be grown and materials can be vermicomposted using low-tech systems, outdoors and year-round.

- **Low temperatures.** *Eisenia* can survive in temperatures as low as 0°C, but they don't reproduce at single-digit temperatures and they don't consume as much food. It is generally considered necessary to keep the temperatures above 10°C (minimum) and preferably 15°C for vermicomposting efficiency and above 15°C (minimum) and preferably 20°C for productive vermiculture operations.
- **High temperatures.** Compost worms can survive temperatures in the mid-30s but prefer a range in the 20s °C. Above 35°C will cause the worms to leave the area. If they cannot leave, they will quickly die. In general, warmer temperatures (above 20°C) stimulate reproduction.
- **Worms's response to temperature differentials.** Compost worms will redistribute themselves within piles, beds or windrows according to temperature gradients. In outdoor composting windrows in wintertime, where internal heat from decomposition is in contrast to frigid external temperatures, the worms will be found in a relatively narrow band at a depth where the temperature is close to optimum. They will also be found in much greater numbers on the south-facing side of windrows in the winter and on the opposite side in the summer.
- **Effects of freezing.** *Eisenia* can survive having their bodies partially encased in frozen bedding and will only die when they are no longer able to consume food.

4.11.5 Other Important Parameters

There are a number of other parameters of importance to vermicomposting and vermiculture:

4.11.5.1 pH: Worms can survive in a pH range of 5 to 9. Most experts feel that the worms prefer a pH of 7 or slightly higher. In general, the pH of worm beds tends to drop over time. If the food sources are alkaline, the effect is a moderating one, tending to neutral or slightly alkaline. If the food source or bedding is acidic (coffee grounds, peat moss) than the pH of the beds can drop well below 7. This can be a problem in terms of the development of pests such as mites. The pH can be adjusted upwards by adding calcium carbonate. In the rare case where they need to be adjusted downwards, acidic bedding such as peat moss can be introduced into the mix.

4.11.5.2 Salt Content: Worms are very sensitive to salts, preferring salt contents less than 0.5%. If saltwater seaweed is used as a feed (and worms do like all forms of seaweed), then it should be rinsed first to wash off the salt left on the surface. Similarly, many types of manure have high soluble salt contents (up to 8%). This is not usually a problem when the manure is used as a feed, because the material is usually applied on top, where the worms can avoid it until the salts are leached out over time by watering or precipitation. If manures are to be used as bedding, they can be leached first to reduce the salt content. This is done by simply running water through the material for a period of time. If the manures are pre-composted outdoors, salts will not be a problem.

4.11.5.3 Urine Content: If the manure is from animals raised or fed off in concrete lots, it will contain excessive urine because the urine cannot drain off into the ground. This manure should be leached before use to remove the urine. Excessive urine will build up dangerous gases in the bedding. The same fact is true of rabbit manure where the manure is dropped on concrete or in pans below the cages.

4.11.5.4 Toxic Components: Different feeds can contain a wide variety of potentially toxic components. Some of the more important are as (a) De-worming medicine in manures, particularly horse manure (b) Detergent cleansers industrial chemicals, pesticides: These can often be found in feeds such as sewage or septic sludge, paper-mill sludge, or some food processing wastes, and (c) Tannins.

4.11.7 Precautions

4.11.7.1 Protection from Pest and Diseases: Flies are commonly attracted to the decomposing organic material. Other than this, the worms are not subjected to diseases caused by micro-organisms. A disease known as "sour crop" caused by environmental conditions and is subjected to predation by insects and certain animals. To avoid the problems, selection of proper bedding material and composting material are must, and also have to maintain moisture and temperature in the composting unit.

4.11.7.2 Protection from sunlight and rain: A thatched roof may be provided to protect the vermicomposting unit from direct sunlight and rain.

4.11.8 Technique to avoid Predators

There are few predators like ants, birds and lizards; they damage or predate the earthworm. To avoid all these problems of predation, the pit or heap should be covered by gunny bags. Make sure that compost beds/heaps are not covered by plastic sheets/material since this can trap the heat and gases due to non-porous nature. The most effective means of controlling ant is to increase the moisture level in bed, so that ants won't be able to tolerate anymore. The composting unit can also be converted into an Island surrounded by water.

4.11.9 Assessing the Maturity of Vermicompost and Harvesting

Black granular compost formation after 40-45 days at the surface of tank indicates the compost is ready for harvesting, which can be done by scrapping layer-wise from top of the tank. Watering should be stopped 5 days before the harvesting and compost should be collected from the top without disturbing the bottom layer.

4.12 Harvesting of Vermicompost

In the tub method of composting, first harvesting can be done after 2 months and the castings formed on the top layer are collected periodically. The collection may be carried out once in a week, scooping the casting with hands and heaped it in a shady place. The harvesting of casting should be restricted to earthworm presence top layer. This periodical harvesting is necessary for free flow of air and retaining the quality of compost. Otherwise, when watering is done the finished compost gets compacted. In case of small bed vermicomposting method, periodical harvesting is not required. Since the height of the waste material heaped is around 1 foot, the produced vermicompost can be harvested at one time after the process is over.

4.12.1 Methods of Earthworm Harvesting

1. Manual method

- Used by small scale growers.
- Involves hand sorting or picking the worm directly from compost by hand.

2. Screen method

- A box is constructed with screen at bottom and compost along with earthworm spread above the box can be separated.

3. Cow Dung Ball method

- A cow dung ball is placed into the bed and the ball is kept for about 24 hrs.
- The cow dung ball should be taken out on the next day and finding all the worms sticking to the ball.
- The worms can be separated out by placing the cow dung ball in a bucket full of water.
- The collected worms can be used for the next batch of composting.

4.12.2 Storing and packing of vermicompost

- The harvested vermicompost should be stored in dark and cool place and it should be protected from sunlight.
- It is more advisable to store the compost in open dark room rather than closed sector.
- The moisture level of prepared compost should be maintained, so packing should be done at the time of selling.
- The compost can be stored for one year without loss of quality, if moisture is maintained at 40 % level.

4.13 Vermiwash

Earthworms play a vital role in plant growth. It is a quite possible to effect quick change over for sustainable agriculture by harnessing brand new vermicompost technology to the soil. Now a days, the commercial vermin culturists have started promoting a product called vermiwash. Vermiwash (clear and transparent, pale yellow coloured fluid). This vermiwash would have enzymes, secretions of earthworms which would stimulate the growth and yield of crops and even develop resistance in crops receiving this spray. Such a preparation would certainly have the soluble plant nutrients apart from some organic acids and mucus of earthworms and microbes (Shivsubramanian and Ganeshkumar, 2004). But so far there are no experimental evidences to quantify the effect of such spray. Microbes in the environment significantly influence the biogeochemical cycle of phosphorus. The organic phosphorous compounds are decomposed and mineralized by enzymatic complexes like phosphatases produced by microbes. In the ecosystem, a mixed population of microbes is essential to promote enzymatic degradation of naturally occurring phosphorous compounds (Trivedi and Bhatt, 2006). Vermiwash are transported to the leaf, shoots and other parts of the plants in the natural ecosystem as foliar spray. Shweta and Singh (2007) reported that presence of plant growth promoting substance in vermicompost and Subasashri (2004) reported vermiwash as an effective biopesticides.

4.13.1 Principle of Vermiwash Production

The basic principle of Vermiwash preparation is simple. Worm converted soils have burrows formed by the earthworms. Bacteria richly inhabit these burrows, also called as the drilospheres. Water passing through these passages washes the nutrients from these burrows to the roots to be absorbed by the plants. This principle is applied in the preparation of vermiwash. Vermiwash can be produced by allowing water to percolate through the tunnels made by the earthworms on the coconut leaf - cow dung substrate kept in a plastic barrel. Water is allowed to fall drop by drop from a pot hung above the barrel into the vermicomposting system. But barrels are not a must for Vermiwash preparation. Vermiwash units can be set up either in barrels or in buckets or even in small earthen pots (Fig. 4.3).

Table 4.4A: Composition of Vermiwash:

1	pH	7.48 ± 0.03
2	Organic carbon%	0.008 ± 0.001
3	Nitrogen%	0.01 ± 0.005
4	Available phosphorous%	1.69 ± 0.05
5	Potassium (ppm)	25 ± 2

Table 4.4B: Micro-nutrients in Vermiwash

Sl. No.	Nutrient	Quantity (ppm)
1	Calcium	3 ± 1
2	Copper	0.01 ± 0.001
3	Ferrous	0.06 ± 0.001
4	Magnesium	158.44 ± 23.42
5	Manganese	0.58 ± 0.040
6	Sodium	8 ± 1
7	Zinc	0.02 ± 0.001

Table 4.4C: Microbial composition of Vermiwash

Sl. No.	Microbes	Microbial count (CFU/ml)
1	Total heterotrophs	1.79 x10^3
2.	Total fungi	1.46 x 10^3
3	Nitrobacter sp.	1.12 x 10^3
4	Nitrosomonas sp.	1.01 x10^3

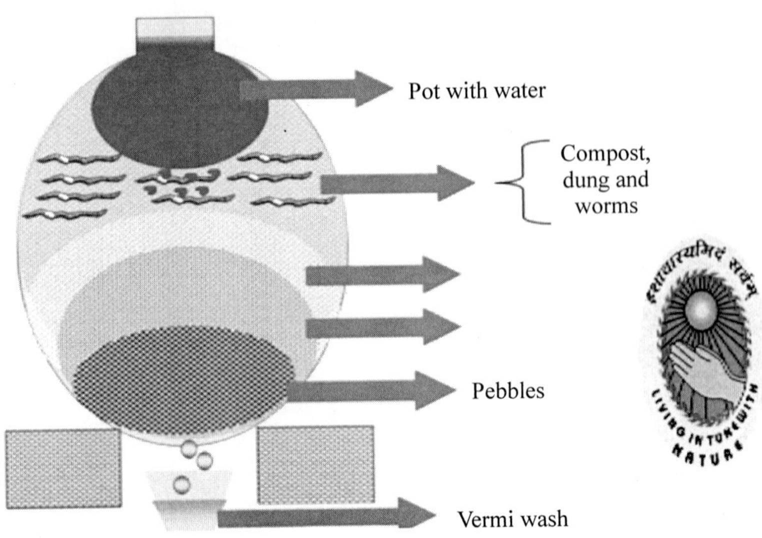

Fig. 4.3 Production of Vermiwash

4.13.2 Method of Preparation of Vermiwash

Vermiwash preparation is completed into following steps such as-

- Select one sufficiently large container made of concrete or plastic bucket.
- Drill a hole on the base of the container to fix a tap to it. A base layer of gravel or broken small pieces of bricks are placed up to height of 10-15cm.
- On the coarse sand layer place 40-45cm pre-decomposed organic wastes and moistens the different layers by using water.
- Introduce about 2000 earthworms into the container. To get vermiwash continuously suspend a mud pot or a small bucket with some holes. Cotton wicks/or bamboo sticks are placed in the holes so that water can trickle down.
- Fill the container with 4-5 liters water every day.
- After 10 days vermiwash starts forming in the container.
- Everyday about 3-4 liters of vermiwash can be collected.

4.13.3 Benefits of Vermiwash

- A mixture of Vermiwash (1litre) with cow urine (1litre) in 10 liters of water acts as bio-pesticides and liquid manure.

- Vermiwash increases the rate of photosynthesis in crop and rate of decomposition of compost, therefore enhance the crop yield.
- Vermiwash increases the number of micro-organisms in the soil, and enhance the resistance to pest and diseases.

4.13.4 Dosage for Use

a) **Root dip/Stem dip:** The seedlings before transplanting are dipped in Vermiwash solution which is diluted 5 times with water for 15-20 minutes and then transplanted. Similarly the cuttings can also be dipped in the solution.

b) **Foliar spray:** Vermiwash is diluted in water 5 times and sprayed on the foliage of crops. It provides the plant with vital nutrients but also helps to control plant disease.

c) **Soil drench:** Vermiwash is diluted 10 times with water and the soil is drenched with the solution to prevent some of the soil borne pathogens.

4.13.5 Precautions

- Do not allow to compact the contents.
- Do not mix un-decomposed materials while, watering, and add any green material.
- The tap should always be kept open to collect the washings.
- The unit starts yielding good quality Vermiwash after ten days.
- Vermiwash should be stored in cool dry place.
- Water should be poured slowly.

4.14 Benefits of Vermicompost

4.14.1 Nutrient Content of Vermicompost

The plant nutrient contents in vermicompost depend on the source of the raw materials and the earthworm species used for composting. A fine vermicompost is rich in macronutrients like nitrogen (N), phosphorus (P) and potash (K), and other secondary and micronutrients as well. Nutrients in vermicompost are present in readily available form and are released within a month of application. It contains nitrogen fixing bacteria like *Azotobacter* sp., *Agrobactericum* sp. and *Rhizobium* sp. and some phosphate solubilizing bacteria (Zambare et al., 2008) and other nutrients as given in Table 4.5)

Table 4.5A: Common available Nutrients in Vermicompost

Nutrients	Amounts
Organic carbon	9.5 – 17.98%
Nitrogen	0.5 – 1.50%
Phosphorous	0.1 – 0.30%
Potassium	0.15 – 0.56%
Sodium	0.06 – 0.30%
Calcium and Magnesium	22.67 to 47.60 meq/100g
Copper	2 – 9.50 mg kg^{-1}
Iron	2 – 9.30 mg kg^{-1}
Zinc	5.70 – 11.50 mg kg^{-1}
Sulphur	128 – 548 mg kg^{-1}

Table 4.5B: Nutrient content (N, P& K) is higher in vermicompost than rural composts.

Element	Vermicompost	Rural compost
N (%)	1.68	0.50 - 1.0
P (%)	1.06	0.40 – 0.80
K (%)	1.57	0.18 – 1.2

Source: Hazarika *et al.,* (2006); Sharma (2002)

4.14.2 Soil Health Improvement and Crop Productivity Enhancement

Vermicompost is simply the excreta of earthworms rich in plant nutrients, plant growth hormones and humus. It is used as an important source of organic matter to the soil as well as soil amendment for sustainable agricultural production. Its application may result in soil health improvement and crop productivity enhancement because of the following reasons:

a) Vermicompost is free flowing, easy to apply, handle and store and does not have bad odour.

b) Vermicompost improves soil structure, texture, aeration, and water holding capacity and prevents soil erosion.

c) Vermicompost contains earthworm cocoons and increases the population and activity of earthworm in the soil.

d) Vermicompost is rich in beneficial micro flora such as N-fixers, P-solubilizers, cellulose decomposing micro-flora, etc. which can improve soil environment.

e) Vermicompost contains earthworm cocoons and increases the population and activity of earthworm in the soil.

f) Vermicompost prevents nutrient losses and increases the use efficiency of chemical fertilizers.

g) Vermicompost is rich in all essential plant nutrients, and provides excellent effect on overall plant growth' encourages the growth of new shoots/leaves and improves the quality and shelf life of the produce.

h) Vermicompost enhances the decomposition of organic matter in soil, and contains valuable vitamins, enzymes and hormones like auxins and gibberellins etc.

i) Immobilized enzymes such as protease, lipase, amylase, cellulase and chitinase present in vermicompost keep on their function of biodegradation of agricultural residues in the soil so that further microbial attack is speeded up

j) It does not have foul odour unlike manures and decaying organic wastes.

k) Superiority of vermicompost over other synthetic growth media are more pronounced in plant nurseries:

- It can be used as rooting medium and for establishment of saplings in nurseries.
- Provides excellent effect on overall plant growth, encourages the growth of new Shoots/leaves and improves the quality and shelf-life of the produce.

4.14.3 Application of Vermicompost for Different Crops

Mode of vermicompost application depends upon the type of crop grown in the field/nursery. It is applied in the tree basin for fruit crops. It should be added in the pot mixture for potted ornamental plants and for raising seedlings. Vermicompost should be used as a component of integrated nutrient management system for better crop production.

Table 4.6: Recommended quantity and time of application of vermicompost for different crops

Crop	Recommended quantity	Time of application
A. Field crop		
Rice	1 ton/acre	After transplanting
Sugarcane	1.5 ton/acre	Last ploughing
Chilli	1 ton/acre	Last ploughing
Groundnut	0.5 ton/acre	Last ploughing
Maize	1 ton/acre	Last ploughing
Turmeric	1 ton/acre	Last ploughing
B. Fruit crop		
Grape	1 ton/acre	June-July
Citrus. Pomegranate, Guava	2 Kg per tree	At planting time and before flowering in 1-2 year old tree

Crop	Recommended quantity	Time of application
Mango	2 kg per tree 5 kg per tree 10 kg per tree 20 kg per tree	At planting time 1-5 year old tree 6-9 year old tree Tree older than 10 years
C. Vegetable crop		
Onion, Garlic, Potato, tomato, bhendi, brinjal, cabbage, cauliflower	1-1.5 ton/ acre	Last ploughing
D. Tree		
Teak	3 kg per tree	At planting time
E. Flowers	100-200 g/sq ft	At planting time and before flowering

Case studies on the effect of vermicompost application on soil and crop productivity Manivannan *et al.* (2009) found that application of vermicompost @ 5 tonnes/ha gave significantly higher result than the application of inorganic fertilizers @ 20:80:40 kg/ha in Frence bean *(Phaseolus vulgaris)* in terms of growth, yield (1.6 times) and quality (protein (1.05 times) and sugar (1.01 times) content in seed) of bean. Vermicompost application also improve the physical, chemical and biological properties of clay loam and sandy loam soils of Sivapuri, Chidambaram, Tamil Nadu. Rajkhowa and his coworker (2017) reported that the integrated use of 50% RDF + VC 2.5 T/ha + lime 4 q/ha, under the hilly ecosystem of NE India, resulted in significantly higher yield of green gram (10 q/ha) and improved the soil organic carbon (2.5 %), bacteria and fungi population and available N, P_2O_5 and K_2O compared to the sole application of recommended dose of fertilizer.

Ansari and Ismali (2012) concluded that vermicompost could be effectively used for the cultivation of many crops and vegetables, which could be a step towards sustainable organic farming. Such technologies in organic waste management would lead to zero waste techno farms without the organic waste being wasted and burned rather than would result in recycling and reutilization of precious organic waste bringing about bioconservation and biovitalization of natural resources. Choudhary and Suresh Kumar (2013) showed that application of vermicompost can increase the production potential of cowpea (*Vigna unguiculata* L., Walp.) in acid soil by improving water retention at field capacity (FC), permanent wilting point (PWP), bulk density (BD), pH, and availability of nitrogen (N), phosphorus (P), and potassium (K), thereby increasing growth and yield attributes of cow pea. Chaudhuri *et al.,* (2016) observed a significant (p\0.05) but gradual increase in density (up to 20 tons/ha/yr) and biomass (up to 30 tons/ha/yr) of earthworms with the increasing amounts of vermicompost application in pineapple plantation. During the second year, average length and width of leaves, number of leaves per plant,

plant girth, fruit weight, fruit yield and fruiting percentage were highest in plot where vermicompost were applied at 20 tons/ha/yr compared to other treated plots and control. The study revealed that pineapple yield was very much related to the particular concentration of vermicompost, beyond the level of which production declines and increase in vegetative growth, fruit weight and fruiting percentage of pineapple are strongly linked with the soil pH, available P, available K, clay content and the earthworm density of soils. Rekha *et al.* (2018) recorded that C. *annum* treated with 50% vermicompost showed significant growth than the plant growth enhancers' *viz*. Gibberellic acid (GA) and Indole acetic acid (IAA) treated plants. Significant improvement in all the parameters like length of shoot, length of inter node, number of leaves and number of branches was observed in plants at the end of 3rd, 4th and 5th weeks of treatment. The findings clearly indicate that vermicompost can be exploited as a potent biofertilizer.

Table 4.7: Effect of nutrient management on soil nutrient status and microbial population

Treatment	pH	Nutrient (kg/ha)			Soil Organic carbon (%)	Bacterial population (CFU x10^6 g-1)	Fungal population (CFU x10^6 g-1)
		N	P_2O_5	K_2O			
Recommended dose of fertilizer (RDF)	4.73	275.9	20.8	121.9	2.2	177	20.3
50% RDF + Vermicompost (2.5t/ha)	4.84	283.5	25.9	138.0	2.5	202.7	28.0
50% RDF + Lime (4q/ha)	4.80	228.8	19.5	130.0	2.3	184.7	22.7
50% RDF + Vermicompost (2.5t/ha) + Lime (4 q/ha)	4.95	290.6	27.2	151.0	2.5	264.7	44.7

Source: Rajkhowa *et al.* 2017

4.15 Vermicompost: An Additional Source of Income

A growing number of individuals and institutions are taking interest in the production of vermicompost utilising earthworm activity. The operational cost of production of vermicompost in a year works out to be around Rs. 4.2/Kg on an average, thus, it is quite profitable to sell the compost at Rs. 10/Kg. As vermicompost can be produced in any non-economic place with shade, high humidity and cool temperature, it has a near to nil opportunity cost and provide a great scope for earning extra income for the farm. Abandoned cattle shed or poultry shed or unused buildings can be used. It can also be produced in open area, by providing low cost thatched roof to protect the process from direct sunlight and rain. A vermicomposting unit can be set up in any convenient scale according to the availability of resources.

a) **Large scale vermicomposting** - commercial vermicomposting facilities may require 0.5 – 0.6 acre to house 6-8 sheds, store house, manager office, water supply system, etc.

b) **Small scale vermicomposting** - farmers can prepare in their own farms in small brick tank, cement tub, wooden boxes, plastic bins or any other containers (except metal) with a drain hole at the bottom.

Apart from the compost, which has a recovery of 75%, the unit will also have as by products i.e. (a) Adult worms for sale, and (b) Vermi-wash – rich in plant nutrients

4.15.1 Financial Aspect of a Vermicomposting Unit

Two types of cost are incurred in setting up and operationalizing a vermicomposting unit viz., capital cost: incurred on development of working shed or vermished, construction of vermi tank, and other tools and implements including sieves and operational cost: incurred on seed stock (earthworms), biodegradable wastes and labour cost. The cost of production for a vermicomposting unit consisting of a single tank with a capacity of 1000 Kg and 75% compost recovery is worked out and presented in the subsequent (Table 4.8).

Benefits

- It is assumed that there will be around 2-3 cycles of production in the first year and 5 – 6 cycles in the subsequent years with a duration of each cycle at around 65-70 days.
- Benefits include the income from sale of vermicompost (@Rs 10 per kg) and adult worms (@ Re 1 per worm) and vermiwash (@Rs 50 per Litre).

Costs

- Vermicomposting could be taken up commercially on any scale starting from 10 MT per annum (TPA) to 1000 TPA and above.
- For a small scale single tank production, the total capital cost of Rs. 3300 is required first time with a subsequent operational cost of Rs. 605.

Table 4.8: Per Kg cost of production per cycle in a year

Sl. No.	Particulars	Units	Quantity	1st Cycle	2nd Cycle	3rd Cycle
1	Operational costs					
	A. Material costs			4300		
	Agricultural wastes and cow dung	Kg	1000	300	300	300
	Earthworms	Nos.	4000	4000		
	B. Labour costs	Rs.		250	250	250
	• Pit filling	MD	0.25	50		
	• Worm separation	MD	0.25	50		
	• Watering	MD	0.25	50		
	• Collection of wastes	MD	0.25	50		
	Sieving	MD	0.25	50		
	C. Interest on working capital	Rs.		455	55	55
	D. Total operational cost (A + B + C)	Rs.		5005	650	650
2	Capital cost					
	Land rent	Rs.				
	Working shed	Rs.		200		
	Vermi tank	Rs.		3000		
	Tools and implements	Rs.		100		
	E. Total capital cost			3300		
	F. Total production costs (D + E)			8305	605	605
	Total vermicompost production	Kg.	750	7500	7500	7500
	Cost of production per Kg	Rs.		11.07	0.8	0.8

Table 4.9: Cost and benefits of a vermicompost unit of dimension 10′x4′x2′ annually

Sl. No	Cost	Amount (Rs.)
1	Total capital cost (Fixed cost)	3300
2	Total operational cost	6215
3	Total cost Benefit	9515
4a	Sale of vermicompost (@ Rs. 10 per Kg)	22500
4b	Sale of worms (@ Re. 1 per worm)	500
4c	Total benefit	23000
5	Net benefit	13,485

4.15.2 Inference From the Financial Aspects

On an average in a year, 2.25 tonnes of vermicompost can be produced from a vermi tank of dimension 10×4×2 feet square, assuming three numbers of harvest in a year. The cost of producing one kg of vermicompost in a year on an average is Rs 4.2. If the cost of the product is Rs 10 per Kg, then a profit of Rs 13,485 can be earned annually, considering at least 500 adult worms can be sold.

4.15.3 Low Demand in the Market

It is advisable to price it competitively. It will help the producer to get good profit even if the price goes low, since the cost of production is quite low.

4.16 Government Initiatives

The government of India and some of the state government are giving financial supporting to the farmers for the production of quality composts from various organic residues through training and through some schemes.

4.16.1 Paramparagat Krishi Vikas Yojana (PKVY): Launched in 2015 with the aim for supporting and promoting organic farming. Funding pattern - 60:40 by the Central and State Governments in Mainland India while the share is 90:10 for North East India.

Approach

- Time frame for implementation is 3 years in a cluster basis.
- In a cluster, 65% is allocated for small and marginal farmers and 30% specifically for women.
- Assistances are provided to the cluster member at a minimal charges of Rs. 5000/unit.

4.16.2 National Mission for Sustainable Agriculture (NMSA): It is one of the eight missions under National Action Plan on Climate Change (NAPCC). It focuses on enhancing agricultural productivity especially in rain-fed areas with a focus on integrated farming, water use efficiency, soil health management and synergizing resource conservation.

Approach

For vermicomposting unit: The mission incurs 50% of the cost subject to a limit Rs. 5000/- per ha and Rs. 10,000/- per beneficiary.

4.16.3 MNREGA: It has expanded its operation for the improvement of agriculture and allied services over the years. Establishment of vermicompost production unit under the scheme has been initiated. Unit cost of 1 unit is Rs. 9000/- with labour and material ratio of 25:75.

4.16.4 Mission for Integrated Development of Horticulture (MIDH): It was launched for the holistic development of horticulture sector in the country during XII plan from 1st April, 2014. It is the integration of the ongoing schemes of National Horticulture Mission, Horticulture Mission for North East & Himalayan States, National Bamboo Mission, National Horticulture Board, Coconut Development Board and Central Institute for Horticulture, Nagaland.

- For a vermicompost unit of size 30′× 8′×2.5′, assistance is provided up-to 50% of the total cost subject to a maximum of Rs. 50,000/- per beneficiary.
- For smaller unit: Size 12′× 4′×2′ up to 50% assistance of cost up to Rs. 8000/-.

(Note: The farmers can also get the financial support for establishing the vermicompost unit from NGOs or by forming the self-help group, consisting 15 or 20 members. Centralized bank like Central Bank of India also offers loans to meet the investment credit or working capital requirement of farmer/s and or corporates for setting up and running Units through schemes like Cent Vermicompost Scheme).

4.17 Pests and Diseases of Vermicompost

Compost worms are not subject to diseases caused by micro-organisms, but they are subject to predation by certain animals and insects (red mites are the worst) and to a disease known as "sour crop" caused by environmental conditions.

- **Ants:** These insects are more of a problem because they consume the feed meant for the worms. Ants are particularly attracted to sugar, so avoiding sweet feeds in the worm beds reduces this problem to a minor one. Keeping the bedding above pH 7 also helps
- **Birds:** They are not usually a major problem, but if they discover your beds they will come around regularly and help themselves to some of your workforce. Putting a window cover of some type over the material will eliminate this problem. These covers are also useful for retaining moisture and preventing too much leaching during rainfall events. Old carpet can be used for this purpose and is very effective.
- **Centipedes:** These insects eat compost worms and their cocoons. Fortunately, they do not seem to multiply to a great extent within worm beds or windrows, so damage is usually light. If they do become a problem, one method suggested for reducing their numbers is to heavily wet (but not quite flood) the worm beds. The water forces centipedes and other insect pests (but not the worms) to the surface, where they can be destroyed by means of a hand-held propane torch or something similar.
- **Mites:** There are a number of different types of mites that appear in vermiculture and vermicomposting operations, but only one type is a serious problem: red mites. White and brown mites compete with worms for food and can thus have some economic impact, but red mites are

parasitic on earthworms. They suck blood or body fluid from worms and they can also suck fluid from cocoons. The best prevention for red mites is to make sure that the pH stays at neutral or above. This can be done by keeping the moisture levels below 85% and through the addition of calcium carbonate, as required.

- **Moles:** Earthworms are moles' natural food, so if a mole gets access to your worm bed, you can lose a lot of worms very quickly. This is usually only a problem when using windrows or other open-air systems in fields. It can be prevented by putting some form of barrier, such as wire mesh, paving, or a good layer of clay, under the windrow.
- **Sour crop or protein poisoning:** This "disease" is actually the result of too much protein in the bedding. This happens when the worms are overfed. Protein builds up in the bedding and produces acids and gases as it decays. According to some workers: "when you see a worm with a swollen clitellum or see one crawling aimlessly around on top of the bedding, you can just bet on sour crop and act accordingly, but fast". The solution is a "massive dose of one of the mycins, such as farmers give to chicken or cattle". Farmers wishing to avoid these or similar antibiotics should work to prevent sour crop by not overfeeding and by monitoring and adjusting pH on a regular basis. Keeping the pH at neutral or above will preclude the need for these measures.

4.18 Precautions

a. Do not overload the vermibed with organic matter because rise to thermophilic temperature will adversely affect their numbers.

b. Do not cover vermicompost system with plastic sheets as it may trap heat and gases.

c. Proper maintenance of moisture is extremely important. Dry conditions kill the worms and waterlogging condition drives them away.

d. Avoid addition of higher quantities of acid rich substances like tomatoes or citrus fruit wastes.

e. Precaution against attacks by red ants who may also feed on cocoons is required by spraying a suitable decoction around the enclosure. A decoction containing 20 litre water, 100g each of chili powder, turmeric powder and salt and a little soap powder may be sprayed around the units on the soil to keep the red ants under control, Neem oil (0.5%) may also be used (Ismail, 2000).

References

Ansari, A.A.; Ismail, S.A. (2001b). Journal of Soil Biology and Ecology 27: 25-27.

Ansari, A.A.; Ismail, S.A. (2008). Pakistan Journal of Agricultural Research 21(1-4): 92-97.

Ansari, A.A.; Ismail, S.A. (2012). Journal of Agricultural Technology 8(2): 403-415.

Ansari, A.A.; S.A. Ismail (2001a). Journal of Soil Biology and Ecology 21:21-24.

Ansari, A.A.; Sukhraj, K. (2010). Pakistan Journal of Agricultural Research 23 (3-4): 137-142.

Bansal, S.; Kapoor, K.K. (2000). Bioresource Technology 73:95-98.

Baskar, A.; Macgregor, A.N.; Krikram, J.H. (1993). Soil Biology and Biochemistry 25:1673-77.

Bezborodov, G.A.; Khalbayeva, R.A. (1990). Soviet Soil Science 22:30-35.

Boruah, T. (2019). Vermitechnology- An overview http://babrone.edu.in/blog/index.php/2019/02/04/vermitechnology-an-overview/

Card, A.B.; Anderson, J.V.; Davis, J.G. (2004). Vermicomposting Horse Manure. Colorado State University Cooperative Extension no. 1.224. Available at http://www.ext.colostate.edu/pubs/livestk/01224.html

Chanu, L.J.; Hazarika, S.; Choudhury, B.U.; Ramesh T.; Balusamy, A.; Moirangthem, P.; Yumnam, A.; Sinha, P.K. (2018). A Guide to vermicomposting-production process and socio economic aspects. Division of Natural Resource management ICAR Research Complex for NEH Region Umroi Road, Umiam, Meghalaya, India-793103.

Chaudhuri, P.S.; Paul, T.K.; Dey, A.; Datta, M., Dey, S.K. (2016). Int. J. Recycl. Org. Waste Agricult 5: 93-103.

Choudhary, V.K.; Suresh Kumar, P. (2013). Commun Soil Sc. Plant Anal. 46(20): 2523-2533.

Darwin, C. (1981). The Formation of Vegetable mould through the action of worms with observations of their habits. Murray, London pp. 298.

Dorcas Joy, E.K.; Reddy, S.M.; Reddy, M.V. (1991). Journal of Soil Biology and Ecology 11:57-61.

Doube, B.M., Schimdt, O., Killham, K.; Correll, R. (1997). Soil. Biol. Biochem., 29: 569-575.

Edwards, C.A.; Lofty, J.R. (1980) Journal of Applied Ecology 17:533-43.

Edwards, C.A., Bohlen, P.J., Linden, D.R.; Subler, S. (1995). Earthworms in agroecosystem. In: Earthworm Ecology and Biogeography in North America. (Hendrix, P. F. eds.), Lewis Publisher, Boca Raton, FL, pp: 185-213.

Gaur, A. C. (2010). In: Biofertilizers in Sustainable Agriculture, ICAR, New Delhi pp. 213-264.

Gaur, A.C; Mathur, R.S. (1990). Soil Fertility and Fertilizers use, Vol 4 (Eds. Kumar *et al.*,) IFFCO, New Delhi.

Gosh, M.; Chattopadhyay, G.N.; Baral, K. (1999). Bioresource Technology 69:149-54.

Heungens, A. (1969). Rev. Ecological, Biological Society 2:131-145.

http://agricoop.nic.in/sites/default/files/Vermicompost%20Production%20Unit.pdf

http://agritech.tnau.ac.in/org_farm/orgfarm_vermicompost.html

Hullegalle, N.R.; Ezumah, H.C. (1991). Agriculture, Ecosystem and Environment 35:55-63.

Ismail, S.A. (2005). The Earthworm Book. Other India Press, Mapusa, Goa. pp. 101..

Iyyanki V. Muralikrishna, Valli Manickam, (2017). Solid Waste Management Environmental Management Science and Engineering for Industry pp. 431-462

Kalpan, D.L.; Hartstein, R.D.; Nauhoser, F.F. (1980). Soil Biology and Biochemistry 12:347-52.

Krishnamoorthy, R.V.; Vajranabhaiah, S.N. (1986). In: Proceedings, Indian Academy of Science (Animal Science) 95:341-51.

Lachnicht, S.L.; Hendrix, P.F. (2001). Spodosol. Soil Biol. Biochem. 33: 1411-1417.

Lee, K.E. (1985). Eartworms, Their Ecology and relationship with Soil and land use. Academic press, New york pp. 441.

Lee, K.E. (1992). Soil Biology and Biochemistry 24:1765-71.

Madsen, E.L.; Alexander, M. (1982). Soil Science Society of American Journal 46:537-60.

Manivannan, S., Balamuruyan, M., Parathasarthi, K., Gunasekaran, G., Ranganathan, L.S. (2009). J. Environ. Biol. 30 (2): 275-81.

OACC Manual of On-Farm Vermicomposting and Vermiculture by Glenn Munroe Organic Agriculture Centre of Canada.

Philip, G. (1981). The University Atlas 21st Edtn., Philip, London, pp. 150.

Rajkhowa, D.J., Sarma, A.K., Mahanta, K., Saikia, U.S.; Krishnappa, R. (2017). India. J. Environ. Biol. 38: 15-19.

Rebatain, S.C.; Stinner, B.R. (1988). Agricultural Ecosystems Environment 24:135-46.

Reddell, F.; Spain, A.V. (1991). Soil Biology and Biochemistry 23:775-78.

Reinecke, A.J. Venter, J. M. (1987). Biological Fertility and Soil 3:135-41.

Reinecke, A.J.; Viljoen, S.A.; Saayman, R.J. (1992). Soil Biology and Biochemistry 24:1295-1307.

Rekha, G.S., Kaleena, P.K., Elumalai, D., Srikumaran, M.P.; Maheswari, V.N. (2018). Int J. Recycl Org Waste Agricult 7(1): 93-103.

Rorat A.; Vandenbulcke, F. (2019). Earthworms converting domestic and food industry wastes into biofertilizer Industrial and Municipal Sludge Emerging Concerns and Scope for Resource Recovery pp. 83-106.

Shipley, A.E. (1970). In: The Cambridge Natural History. (Harmer, S. F. and Shipley, A. E. eds.). Codicote, England.

Shivsubramanian, K.; Ganeshkumar, M. (2004). Madras Agricultural Journal. 91: 221-225.

Shuster, W.D., Subler, S.; McCoy, E.L. (2000). Soil and Tillage Research. 54: 179-189.

Shweta, K.K.; Singh, Y.P. (2006). Asian Journal of Microbiology, Biotechnology and Environmental Science. 8: 831-834.

Stephens, P.M.; Davoren, C.W.; Double, B.M, Ryder, M.H. (1994). Biological Fertilizers and Soils 18:150-54.

Stewart, V.I., Scullion, J. (1988). In: Earthworms in Water and Environment Management (Eds c.A. Edwards and F. Neuhauser). The Hauge, pp. 263-72.

Subasashri M. (2004). Vermiwash an effective biopesticide. The Hindu Newspaper, 30th September, In: Science and Technology Section.

Tomati, U.; Grappelli, A. and Galli, E. (1988). Biological Fertilizers and Soil 5:288-94.

Tomlin, A.D., Shipitalo, M.J., Edwards, W.M.; Protz, R. (1995). Earthworms and their influence on soil structure and infilteration. In: Earthworm ecology and biogeography in North America. (Hendrix, P.F. ed.), Lewis Publishers, Chelsea. pp. 159-184.

Trivedi R.; Bhatt S.A. (2006). Asian Journal of Microbiology, Biotechnology and Environmental Science. 8: 303-305

Viljoen, S. A.; Reinecke, A. R. (1989). S. Afr. J. Zool. 24: 27–32.

Wollters, J.S.; Joergensen, R.G. (1992). Soil, Biology and Biochemistry 24:171-77.

Yeates, G.W. (1981). Pedobiologia. 22: 191-195.

Zambare V.P., Padul M.V., Yadav A.A.; Shete T.B. (2008). ARPN Journal of Agricultural and Biological Science 3(4): 1-5.

Zhang, H.; Schrader, S. (1993). Biological Fertilizer and Soil 15: 229-34.

5

Integrated Agricultural Management Systems

Agriculture has been very successful in addressing the food and fiber needs of today's world population. However, there are increasing concerns about the economic, environmental and social costs of this success. Integrated agricultural systems may provide a means to address these concerns while increasing sustainability. Integrated agricultural systems have multiple enterprises that interact in space and time, resulting in a synergistic resource transfer among enterprises. In an integrated agricultural system, management decisions, such as type and amount of commodities to produce, are predetermined. Full integration of agricultural systems at the producer or community scale may help in slowing or reversing some of the detrimental environmental and economic problems associated with specialized industrial agriculture. Modern agriculture requires intensive inputs. However, the use of forages and other diverse crops in the crop rotation can reduce intensive inputs, while in some cases increasing crop yield, enhancing nutrient cycling, reducing plant disease, and improving soil quality (Krall *et al.,* 1996; Entz *et al.,* (1995, 2002); Schiere *et al.,* 2002; John, 2008). In view of the fact that increased use of pesticides has been drastically endangering the environmental sustainability, integrated approach to pest management needs adequate importance to make the agriculture ecofriendly. Integrated Pest Management (IPM) also makes use of an ideal combination of physical, chemical, biological and cultural methods to contain pest damage with minimum ecological implication. The multifarious harmful consequences of indiscriminate use of pesticides have cropped up as a serious threat to the ecosystem. Plant protection (against pests, diseases and weeds) determines the effectiveness of other inputs in crop production (Ali *et al.,* 2019). Exclusive reliance on pesticides, fungicides and herbicides resulted in pesticide and herbicide resistance, pest resurgence, residues and environmental pollution. This led to the development of integrated plant protection strategies, which are components of sustainable agriculture with a sound ecological foundation. Integrated plant protection should be understood as an ideal combination of agronomical, biological, chemical, physical and other methods of plant protection against entire complex of pests, diseases and weeds in a specified farming ecosystem, with the object of bringing down their infestation to economically insignificant levels with minimum interference on the activity of natural beneficial organisms. The essence of integrated plant protection concept lies in the harmonious integration of compatible multiple methods use singly or in combination against insect pests, pathogens and weeds. Integration of livestock and cropping systems has the potential benefits of enhancing nutrient cycling efficiency, adding value to grain crops, and providing a use for forages and crop residue (Brummer, 1998).

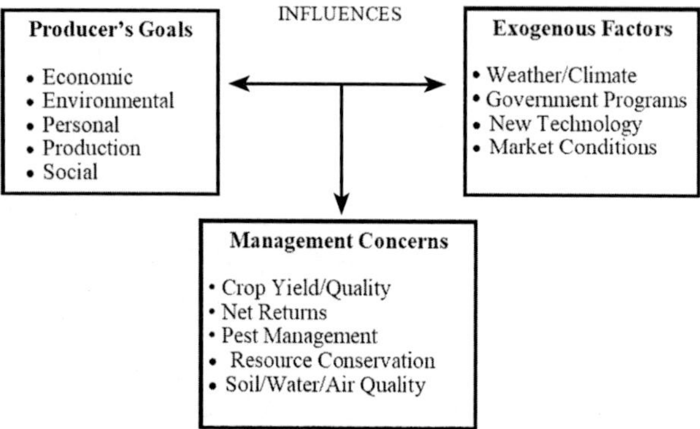

Fig 5.1: Principle of Integrated Agriculture System

5.1 Integrated Disease Management (IDM)

For mitigating the loses due to diseases, several methods such as fungicides, organ mercurial, chemotherapy, thermotherapy, cultural methods and host resistance are employed. However, no single method is effective in controlling a disease. Therefore, IDM became imperative for effective disease control. IDM in organic farming combines the use of various measures. The usefulness of certain measures depends on the specific crop-pathogen combination. In many crops, preventative measures can control diseases without the need of plant protection products. However, for certain disease problems, preventative measures are not sufficient. For example, organic apple production strongly depends on the multiple uses of plant protection products. For many other diseases the role of host resistance, cultural methods and chemical methods are integrated. Solar heat therapy (drying the seed in hot sun after harvest and again before sowing) is a common practice in our agriculture. Among mechanical methods for prevention and against spread of diseases uproot and burn is the age old and the best method so far. It is better to prevent and control vectors against spread of diseases. Disease affected plants are to be uprooted and burnt and alternate and collateral host-crops, grasses, stubbles destroyed. Disease management in organic cropping systems combines various components which can be divided into strategic preventative measures, tactical preventative measures and control measures. For each crop-pathogen relationship and cropping system such components will contribute to different extent to disease management (Termorshuizen, 2002).

5.1.1 Pathogen Characteristics and Disease Management

Host-specificity and mobility are the two main characteristics of pathogens determining the choice of disease management measures (Wijnands *et al.,* 2000). Strictly host-specific pathogens which are not mobile can be controlled by using cropping systems with low frequencies of the susceptible crop. Examples are cyst nematodes of potato or sugar beet. Pathogens which are not host-specific and not mobile can be controlled by the choice and sequence of crops grown in the rotation supported by preventative measures increasing soil suppressiveness and plant health. Examples are the soil borne pathogens *Sclerotinia sclerotiorum* and *Rhizoctonia solani*. Host-specific pathogens with high mobility such as *Phytophthora infestans* in potato cannot be controlled by crop rotation. Preventative measures are sanitation in a cropping area and the choice of crop structure and planting date in combination with resistant varieties. In many situations also control measures such as applications of plant protection products may be needed to achieve sufficient yield. Also pathogens which are not host-specific but highly mobile cannot be controlled by crop rotation. Disease prevention depends on strengthening the crop, escaping the disease by choosing proper seeding dates and creating an open crop structure. Disease control by using crop protection products may be needed in many cases. Example for a mobile pathogen with a broad host range is *Botrytis cinerea* causing grey mould in various crops such as beans, peas, strawberries, grapes and many other crops.

5.1.2 Tactical Preventative Measures

Tactical preventative measures deal with the planning and realization of a certain crop. Typical measures are the choice of variety, seed quality, seeding time and crop structure.

5.1.2.1 Resistant Varieties: In wheat, resistance breeding made considerable progress and partly resistant cultivars are used in practice. In apple, partly resistant varieties are available. However, the pathogen has the potential to adapt. Furthermore, changing varieties in a perennial crop needs high investments.

5.1.2.2 Removal of Crop Residues: *Fusarium* sps. threatening wheat crops are surviving primarily in stubble of cereals including maize. Removing this potential inoculum sources is not feasible, although physical removal especially of maize stubble may have a significant impact on disease development. In apple, removal of fallen leaves as the principle inoculum source of apple scab in spring is an interesting option. Removal of leaves by using specially designed vacuum cleaners has been demonstrated. However, mechanization is difficult, cost intensive and application depends much on orchard circumstances.

5.1.2.3 Biological crop residue treatments: Microbial decomposition of crop residues is a natural process which can be supported by adding stimulating nutrients or selected micro-organisms. Also earthworms can be protected and stimulated to consume plant residues on the soil surface. In arable crops, stimulation of resident microbial populations on residues may be achieved by creating a suitable microclimate, e.g. by using mulches.

5.1.2.4 Healthy seeds and planting material: Seeds of wheat can be infected by *Fusarium* spp. Producing healthy seeds is important to guarantee the establishment of a vigorous crop. For the development of FHB epidemics after flowering, the major inoculum sources are infested crop residues and thus field-borne. Reducing the seed-borne fraction of the disease inoculum may only have very limited effect against FHB. Using clean planting material of apple will not result in any disease prevention since *V. inaequalis* overwinters on the orchard floor and easily can enter disease–free young trees.

5.1.2.5 Sowing time: For infections of wheat ears by *Fusarium* sps., the crucial factor are the climatic conditions during the short window of flowering. Choosing early or late sowing times is not an option for disease prevention since weather during flowering cannot be predicted. Also for apple, no effect of planting time on apple scab can be expected.

5.1.2.6 Crop Structure: Crop structure affects microclimatic conditions within the canopy and determines the distance pathogen spores have to spread to reach susceptible host tissue. A dense wheat crop will favour pathogen sporulation on the soil, but may block spore flights of *Fusarium* sp. depending mainly on splash dispersal during rainfalls, vertical leaf positions may also block spore flights. The canopy structure of apple trees is managed to obtain sufficient yield and possibilities to create more open canopies are limited. Spores of *V. inaequalis* are very much adapted to infect trees and spread from orchards floors and within canopies. Possibilities to manage the apple scab by crop structure are low.

5.1.3 Strategic Preventative Measures

Many measures for preventative disease control have a long-term strategic character. Various aspects of the farm management and the long-term cropping system as well as of the location including the farm neighbourhood have impact on diseases of crops and thus should generally be considered in integrated disease management.

5.1.3.1 Soil Properties: Soil structure, soil suppressiveness, biological soil disinfection, and catch crops are important for managing soil borne diseases but will have no direct effect on the above-ground development of *Fusarium* spp. and *V. inaequalis*.

5.1.3.2 Avoidance of pathogen sources in neighbourhood of field and crop rotation in neighbouring field: Since most *Fusarium* spores travel only a few centimeters, sources in the crop neighbourhood will not cause epidemics. Ascospores of *V. inaequalis* produced in neighbouring orchards may reach the crop. Abandoned orchards and orchards with high apple scab pressure should not be found in the neighbourhood of an apple orchard.

5.1.3.3 Avoiding growth of maize in rotation with wheat will substantially reduce risks of FHB epidemics. Rotation schemes with cereals grown after cereals should generally be avoided. In the perennial apple production crop rotation is no issue.

5.1.3.4 Tillage: Primary inoculum of Fusarium head blight (FHB) are crop residues left on the soil after tillage. Using reduced tillage systems will increase FHB risks since much more residues will be present on the soil surface. In apple orchards, tillage is not an option.

5.1.3.5 Crop rotation: Main inoculum source of FHB are crop residues of preceding diseased crops. The best documented example is the high risk of FHB when wheat is grown after maize. Maize stubble are often colonized by the same *Fusarium* sp affecting wheat and such *Fusarium* sp can survive and multiply on maize stubble for several years.

5.1.4 Disease Control Measures

Disease control measures are used to control a certain disease of a crop. Physical, chemical or biological control measures may be used. Physical treatments. *Fusarium* spp. on seeds can be controlled by warm water treatments. However, the effect on FHB will be limited. Natural compounds and biocontrol agents as plant protection products. The control of FHB does not depend on plant protection products since preventative measures such as rotation, and tillage can be used. In apple, preventative measures such as removal of fallen leaves can delay the outbreak of epidemics. However, epidemics need to be controlled by multiple applications of plant protection products such as copper. Environmentally friendly new products are strongly needed. Biological based technologies are most effective when integrated with physical and chemical approaches in the process towards sustainable ecological based plant diseases management. The important measures are:

a) Use of 2, 4-Diacetyl fluoroglucinol for all diseases of wheat.

b) Use of antagonists is equivalent to natural enemies used in control of insects. Seeds treated with antagonists can be used as feed and food.

c) Use of Biosave-10 and Biosave-11 (based on strains of Pseudomonas syringal against storage rot of vegetables and fruits.

d) Use of molecular tools to assure pathogen free planting materials.

e) Use of non-pathogenic materials like F. oxysporum (Fusariclean) and F. fluorescens (Biocoat) for Fusarium wilt in vegetables and flower crops.

f) Use of root-knot nematode.

5.2 Integrated Pest Management (IPM)

Integrated pest management (IPM), which by definition is a pest management system that, in the context of associated environment and population dynamics, utilizes all the appropriate techniques to minimize the pest population at levels below those causing economic injury. The concept of IPM, a sustainable strategy for managing pests, has been in practice for a long time. Although multiple sources define IPM in different ways, previous models primarily focused on the ecological, and to some extent on the evolutionary, aspects of pest management (Peterson *et al.,* 2018). A recent IPM pyramid presented by Stenberg (2017) identified a lack of a holistic IPM approach that uses both traditional and modern tools. However, his conceptual framework mainly dealt with the ecological aspects of pest management with an emphasis on interdisciplinary research approach. Several reports indicated that IPM implementation depends on numerous factors including the level of education, economic and social conditions, environmental awareness, rational thinking, moral values, regulatory aspects, government policies, availability of IPM tools, extension education, consumer preference, and retail marketing (Parsa *et al.,* 2014; Lefebvre *et al.,* 2015; Jayasooriya and Aheeyar, 2016; Rezaei *et al.,* 2019). However, there is no IPM model that encompasses all these factors and provides a comprehensive description. The more that you know about a pest, the easier and more successful pest management becomes. Once you have identified a pest, you can access information about its life cycle and behavior, the factors that favor development, and the recommended control procedures. Following identification of an insect pest is monitoring to determine the pest status. If there is a need to control the pest, based on monitoring, then you develop a management program followed by implementation and evaluation as illustrated in Fig. 5.2.

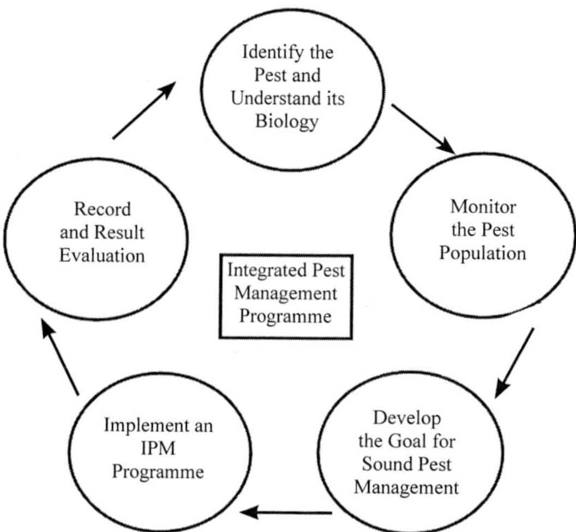

Fig. 5.2: Components of IPM programme

5.2.1 Agronomic Practices for IPM

Adopting good agronomic practices that avoid or reduce pest infestations and damage refers to cultural control.

5.2.1.1 Ploughing and Sanitation: Plowing is also an important control option to destroy the crop residue and expose the soil-inhabiting stages of several vegetable pests (Kunjwal and Srivastava, 2018). Sanitation practices to remove infected/infested plant material, regular cleaning field equipment, avoiding accidental contamination of healthy fields through human activity are also important to prevent the pest spread. Sanitation practices such as bagging unmarketable berries or even changing the harvest schedule from every 3 days to one or two days reduced spotted-wing drosophila (*Drosophila suzukii*) infestations reported by Leach *et al.,* (2017).

5.2.1.2 Selection of Cultivars: Cultivars with high yield potential and quality without resistance to pests and diseases are the main causes of frequent epidemics and mass multiplication of pests and diseases. A large number of cultivars resistance/tolerance to pest and diseases has been developed to suit different agro-ecosystems. Selection of such cultivars can bring down the losses considerably.

5.2.1.3 Manures and Fertilizers: Excessive nitrogen increases susceptibility of crop to sucking and leaf eating pests. Higher rates of nitrogen application than the recommended rate to hybrids without corresponding increase in phosphorus and potassium is the main factor for heavy pest and disease incidence. Increased plant nitrogen can exacerbate arthropod infestations (Hodson and Lampinen, 2018). High (Mitchell *et al.,* 2003) or low nitrogen (Snoeijers *et al.,* 2000) content in the plant can also contribute to some disease problems. Balanced application of NPK helps the crop to tolerate pests and diseases considerably. Low potassium content in plants induces jasmonic acid synthesis in plants and helps with plant's ability to withstand insect pests and certain diseases (Davis *et al.,* 2018).

5.2.1.4 Time of Sowing: As weather influences developmental rhythm of plants as well as growth and survival of pests and diseases, serious setback occurs when the weather conditions are such as to bring about coincidence of susceptible growth stages with highest incidence of pests and diseases. Therefore, adjustment in sowing dates is often resorted to as an agronomic strategy to minimize the crop losses. Maize sown late suffers little borer damage, as by then the egg parasite *Trichogramma* is able to keep down the population of the pest. Rice may suffer less from borer attack if planted early (early June). Early maturing cotton cultivars have become popular in Punjab and Haryana as they escape pink bollworm. Adjusting planting dates can help escape pest occurrence or avoid most vulnerable stages. Early planting of cowpea reduced aphid, thrips, and pod bug infestations in Uganda (Karungi *et al.,* 2000) and the legume podborer (*Maruca vitrata*), the legume flower thrips (*Megalurothrips sjostedti*), and the pod sucking bug (*Clarigralla tomentosicollis*) in Nigeria (Asante *et al.,* 2001).

5.2.1.5 Plant Population: Plant population per unit area influence crop microclimate and resulted impact on pest infestations. Dense plant canopy leads to high humidity build up congenial for pest and disease multiplication. High plant density reduced root maggot (Delia spp.) infestations in canola in Canada (Dosdall *et al.,* 1996) and aphid infestations in cowpea in Uganda (Karungi *et al.,* 2000).

5.2.1.6 Irrigation Management: Irrigation can reduce the soil inhabiting pests by suffocation or exposing them to soil surface to be preyed upon by birds. Irrigating potato crop at tuber formation can minimize potato scab. Anthracnose of beans, early blight and charcoal rot of potato can be checked by furrow irrigation than sprinkler irrigation.

5.2.1.7 Crop Rotation and Intercropping/Habitat Diversification: Many pests prefer feeding on a particular plant or others. This preference may be exploited to reduce the pest load on crop. Crop rotation with non-host or tolerant crops will break the pest cycles and reduce their buildup year after year. Crop rotation tactic has been used for insect, disease, and weed management in many cropping systems (Curl, 1963; Wright, 1984; Liebman and Dyck, 1993; Mohler and Johnson, 2009). Intercropping of non-host plants or those that deter pests or using trap crops to divert pests away from the main crop are some of the other cultural control strategies in IPM (Pretty and Bharucha, 2015; Nielsen *et al.,* 2016).

5.2.2 Methods for IPM

5.2.2.1 Physical or Mechanical Control

This approach refers to the use of a variety of physical or mechanical techniques for pest exclusion, trapping (in some cases similar to the behavioral control), removal, or destruction (Webb and Linda, 1992, Gamliel and Katan, 2012; Gogo *et al.,* 2014; Dara *et al.,* 2018; Dara, 2019). Pest exclusion with netting or row covers, handpicking or vacuuming to remove pests, mechanical tools for weed control, traps for rodent pests, modifying environmental conditions such as heat or humidity in greenhouses, steam sterilization or solarization, visual or physical bird deterrents such as reflective material or sonic devices are some examples of physical or mechanical control.

5.2.2.2 Trap Cropping

Trap cropping is the planting of a trap crop to protect the main cash crop from a certain pest or several pests. The trap crop can be from the same or different family group, than that of the main crop, as long as it is more attractive to the pest. There are two types of planting the trap crops; perimeter trap cropping and row intercropping. Perimeter trap cropping (border trap cropping) is the planting of trap crop completely surrounding the main cash crop. It prevents a pest attack that comes from all sides of the field. It works best on pests that are found near the borderline of the farm. Row intercropping is the planting of the trap crop in alternating rows within the main crop. Trap cropping has several advantages i.e. (a) lessens the use of pesticide (b) lowers the pesticide cost (c) preserves the indigenous natural enemies (d) improves the crop's quality and (e) helps conserve the soil and the environment.

5.2.2.3 Bird Perches

Bird perches are resting places for predatory birds to rest and to look for preys; such as insect pests of cotton, peanuts, and cowpeas. Predatory birds prefer to look for prey in field crops where they have places to rest. These plants are found to be attractive to predatory birds. The birds feed on their seeds. In cotton field, plant Setaria in every 9th or 10th row of cotton. Once the birds are on the field, they prey on cotton bollworms and other insects. To make bird perches, use bamboo or wooden poles or tree branches. Erect either of these at regular intervals in the field.

5.2.2.4 Behavioral Methods

The behavior of the pest can be exploited for its monitoring and control through baits, traps, and mating disruption techniques (Heinz *et al.,* 1992, Shorey and Gerber, 1996; Foster and Harris 1997; Vladés *et al.,* 2005; El-Sayed *et al.,* 2009; Morrison *et al.,* 2016). Baits containing poisonous material will attract and kill the pests when distributed in the field or placed in traps. Pests are attracted to certain colors, lights, odors of attractants or pheromones. Behavioural methods are witnessed under pheromones and fairomones.

5.2.4.1 Pheromones: Pheromones are ectohormones secreted by an organism, which elicit behavioural responses from other members/sex of its own species. These are extremely selective, nontoxic, highly biodegradable and effective at low application rates. Synthetic sex pheromones are commercially available and are used for surveillance, monitoring and control of many lepidopterous pests such as spotted bollworm, tobacco caterpillar, potato tuber moth, diamond back moth and leaf folder etc. Pheromone lures confuse adult insects and disrupt their mating potential, and thus reduce their offspring.

5.2.4.2 Fairomones: These are volatile compounds that evoke behavioural response adaptively favourable to the receiver. Fairomones are released either by the host plant or by the host insects. While former issued by the pest and natural enemies to locate their habitats, later is used for prey finding and parasitization/preying. Fairomones from host plant can be effectively used to mass trap pest species as well as for monitoring. Use of fairomonal compounds to increase the efficiency of the predator *C. carnea* and the egg parasitoid *T. chilonis* had been successfully demonstrated.

5.2.5 Host Plant Resistance

A strategy that involves the use of pest-resistant and pest-tolerant cultivars developed through traditional breeding or genetic engineering (Douglas, 2018). These cultivars possess physical, morphological, or biochemical characters that reduce the plant's attractiveness or suitability for the pest to feed, develop, or reproduce successfully. These cultivars resist or tolerate pest damage and thus reduce the yield losses. This option is the first line of defense in IPM.

5.2.6 Chemical Control

Chemical control typically refers to the use of synthetic chemical pesticides (Pimental, 2009). However, to be technically accurate, chemical control should include synthetic chemicals as well as chemicals of microbial or botanical origin. Although botanical extracts such as azadirachtin and pyrethrins, and microbe-derived toxic metabolites such as avermectin and spinosad are regarded as biologicals (Lasota and Dybas, 1991; Sarfraz *et al.,* 2005; Dodia *et al.,* 2010), they are still chemical molecules, similar to synthetic chemicals, and possess many of the human and environmental safety risks as chemical pesticides. Chemical pesticides are categorized into different groups based on their mode of action (IRAC 2018) and rotating chemicals from different mode of action groups is recommended to reduce the risk of resistance development (Sparks and Nauen, 2015). Government regulations restrict the time and amount of certain chemical pesticides and help mitigate the associated risks. Certain biostimulants based on minerals, microbes, plant extracts, seaweed or algae impart induced systemic resistance to pests, diseases, and abiotic stressors, but are applied as amendments without any claims for pest or disease control (Larkin 2008; Sharma *et al.,* 2014; López- Bucio *et al.,* 2015; Dara 2018). These new products or technologies can fall into one or more abovementioned categories of pest management. All the pest management options need careful consideration and application to avoid potential risks. For example, several pests developed resistance to transgenic crops with *Bacillus thuringiensis* toxic proteins (Tabashnik *et al.,* 2013) and planting non-transgenic plants along with resistant plants has been recommend, among other strategies, to reduce the resistance development (Tang *et al.,* 2001, Huang *et al.,* 2011). Pests can also develop resistance to botanical and microbial pesticides if they are overused (Dara, 2017). Although pest management decisions are supposed to be based on economic injury levels and thresholds, in many situations they are either not available, difficult to determine, not applicable to all geographic regions or seasons, or existing ones need revalidation (Poston *et al.,* 1983). Some of the established thresholds are also questionable. Because crop production is highly precise due to modern technologies on one hand, as well as highly variable depending on a myriad of biotic and abiotic factors and the proprietary practices of different farming operations, information management and decision-making parts play a critical role in IPM. One cannot offer a one-size-fits-all solution and the pest control efficacy depends on several factors in addition to the option used.

5.2.7 Biological Methods

Biological control basically means "the utilization of any living organism for the control of insect-pest, diseases and weeds". Biological control of insect-pests is gaining recognition as an essential component of successful IPM. Biological control agents for insect pests are available in nature abundantly and work against crop pests naturally called as natural control. The pest management programme where natural enemies of the crop pests form the core component is designated as Bio-Intensive Pest Management (BIPM). Natural enemies such as predatory arthropods and parasitic wasps can be very effective in causing significant reductions in pest populations in certain circumstances

(Hajek and Eilenberg, 2018). Periodical releases of commercially available natural enemies or conserving natural enemy populations by providing refuges or avoiding practices that harm them are some of the common practices to control endemic pests. Biological control has been successfully used in greenhouses (van Lenteren, 1988) and specialty crops such as strawberries grown in the field (Zalom *et al.,* 2018). To address invasive pest issues, classical biological control approach is typically used where natural enemies from the native region of the invasive pest are imported, multiplied, and released in the new habitat of the pest (Kenis *et al.,* 2017; Heimpel and Cock, 2018). The release of irradiated, sterile insects is another biological control technique that has been effectively used against a number of pests (Klassen and Curtis, 2005). The most commonly used bio-agents in BIPM are broadly classified into 3 categories i.e. a) Parasitoid b) Predators, and c) Microbes causes diseases in insects.

5.2.7.1 Parasitoid

These are the insects, either equal or lesser than the size of the host insect (pest) and always require passing at least one stage of their life cycle inside the host system. Due to their high multiplication rates they are of vital importance in the biological control of insect-pests. Insect parasitoids (parasites thriving on insects) include *Trichogamma, Bracon hebetor, Elasmus, Eribor, Trichospilus, Gonlozus, Tetrastichus* and *Chelonus* etc.

5.2.7.2 Predators

These are those insects, which are generally bigger than the host and feed on several of the pests by predating upon them externally. They will be consuming several of the insect-pests during their life cycle and hold a key role in minimizing pest population in field conditions. Commonly used predators are *Chilizonus, Cryptolaemus, Scymnus, Meoichlus* and *Pharoscymnus*. These predators feed on mealy bugs, coccids, scales and mites on citrus, grapevine and guava. Mirid predator *Crytorhinus* attack brown plant hopper of rice. Insect predator Chrysopa is effective on aphids, mealy bugs and young caterpillars. *Chrysoperla* spp. prey upon several of the soft bodied insects (such as aphids, leaf hoppers etc) ladybird beetle against aphids and mealy bugs; spiders against a varied number and types of insects especially in rice ecosystem.

5.2.7.3 Microbes

These are those micro-organism which are capable of causing diseases in insects as a result they lose their appetite, subjected to several physiological disturbances, leading to the ultimate death of thc insect. The microorganisms exploited in biological control of insect-pests are broadly classified into a)

Insect b) viruses c) bacteria d) Entomopathogenic fungi e) Entomopathogenic nematodes and (f) other organisms (i.e. Protozoans and Rickettesia etc). while several antagonistic fungi and bacteria are being successfully used in minimizing the plant disease incidence. Nematode pest management by using biotic agents is also one of the most promising areas and gaining much desired importance in the current scenario of organic farming. Using entomopathogenic bacteria, fungi, microsporidia, nematodes, or viruses, and fermentation byproducts of some microbes against arthropod pests, plant parasitic nematodes, and plant pathogens generally come under microbial control (Mankau, 1981; Paulitz and Bélanger, 2001; Dong and Zhang 2006; Lacey 2017). Pathogens causing diseases in insects and destroying them as:

- **Nuclear Polyhedroses Viruses (NPV):** They viruses cause typical "tree top disease" in insects. The infected insect loose appetite becomes restless and reaches the apical portion of the plant due to O2 depletion inside the insect system. Later the infected insect will be dying by hanging itself in an inverted 'V' shape on the apical portion of the plant. Effective against lepidopteran insects in different crops. *Ha* NPV is used for the management of *Helicoverpa armigera* while *Sl* NPV is meant for *Spodoptera litura*. Similarly, castor semilooper is managed by *Ach* NPV and red hairy caterpillar by *Am* NPV.
- **Granulosis Viruses (GV) and Cytoplasmic Viruses (CPV):** Extensively used against insect-pests of sugarcane. Entomopathogenic viruses are highly specific to host insects which make them exceptionally safe to non-target organisms and nature.
- ***Bacillus thuringiensis*:** This bacterium is highly effective against several insect-pests belonging to order Lepidoptera. They cause disease due to which insect turns black and die. Fungi *Metarhizium anisopliae*, *Beauveria bassiana* and *Verticillium lecanii* are used against important pests like gram pod borer, tobacco caterpillar and sucking pests like thrips, aphids and mealy bugs. The fungi develop hyphae inside insect system resulting in the death of the insect due to mechanical congestion. The mode of action makes these organisms to perfectly suit to the needs of organic farming. In certain cases they produce toxins to kill the insect.
- ***Heterorhabditi* sp and *Steimernema* sp:** These nematodes harbor certain bacteria which act as toxins to insect systems.
- ***Variomorpha* sp:** These protozoa's were found effective against insect pests and can effectively be incorporated as tools in organic farming.

5.2.8 Bio-pesticides

Natural occurrence of diseases caused by micro-organisms is common in both insects and weeds and is a major natural mortality factor in most situations. Use of micro-organisms for pest control involves their culture in artificial media and later introduction of larger amounts of inoculums in to the field at appropriate time. Many fungi and bacteria can be handled in this way but insect viruses have the limitation that they have to be raised in living insects. As the biocontrol agents (microbial pathogens) are applied on the targeted pests in much the same way as chemical pesticides, they are often termed as bio-pesticides or natural pesticides. *Bacillus thuringensis*, a bacterial pathogen infesting a wide range of insect pests, is the most common microbial insecticide in use today. It is used against caterpillars that attack a wide range of crop. Unlike most other chemical insecticides, it can be used on edible products up to the time of harvest. It is selective in action and does not harm parasites, predators or pests. The bacteria come in several commercial formulations such as Dipel, Delfin, Halt, Spicturin, Biolep, BioAsp etc. Another bacterium *B popilaleis* also commonly available against white grub *Popillae japonica* and *Hototricha* sp. Amongst insect pathogenic fungi, commercial preparations of *Verticillium lecanii* are available for the control of aphids, thrips and white fly under glass house conditions.

5.2.9 Botanicals

Some weeds (Lantana, Notchi, Tulsi, Adathoda etc) act as natural repellant to many pests. Trees like pungam, wood apple, anona and their byproducts have excellent insecticidal value in controlling diamond back moth, heliothis, white flies, leafhoppers and aphid infestation. Most commonly used botanicals are neem (*Azadirachta indica*), pungamia (*Pungamia glabra*) and mahua (*Madhuca indica*). Neem seed kernel extract (2 to 5%) has been found effective against several pests including rice cutworm, diamond back moth, rice BPH, rice GLH, tobacco caterpillar, aphids and mites. The pesticidal ingredients of neem formulations belonging to general class of natural products called triterpenes, more specifically, limonides. They act as repellents and also disrupt growth and reproduction in insects. The efficiency of vegetable oils in preventing infestation of stored product pests such as bruchids, rice and maize weevils has been well documented. Root extracts of asparagus work as a nematicide for plant parasitic nematodes. Similarly leaf extracts of many plants can inhibit a number of fungal pathogens.

5.3 Integrated Weed Management (IWM)

IWM involves the concept of multiple tactics of weed management, maintenance of weed population below economic injury level and conservation of environmental quality. A successful IWM strategy has the principle of enhancing farmers' profitability, environmental protection and responsiveness to consumer preference. Weeds vary so much in their growth habit and life cycle under different ecosystems and growing seasons that no single method of weed management can provide effective weed control. Continuous use of one method of weed control creates problems of buildup of weeds that are tolerant to that particular method of weed control. Similarly, shift in weed flora from annual grasses to sedges and appearance of resistant biotypes due to continuous use of some herbicides has been reported. Thus, Integrated weed management is defined in a range of ways, but, at its core, is the idea that many different weed management tools be used, in an integrated way, to manage weeds.

5.3.1 Conceptualization of Integrated Weed Management

One way to conceptualize integrated weed management is overlapping the four sciences/means of managing weeds: physics, chemistry, biology, and ecology (Merfield, 2019)

a) **Physical weed management** approaches include mechanical techniques such as hoeing and tillage and thermal techniques such as flame weeding;

b) **Chemical weed management** is dominated by the synthetic herbicides, but, there are also natural herbicides;

c) **Biological weed management** uses an understanding of plant biology, for example germination, to manage weeds;

d) **Ecological weed management** uses the interactions among species to achieve weed control, e.g., crop-weed competition, allelopathy, and crop rotations.

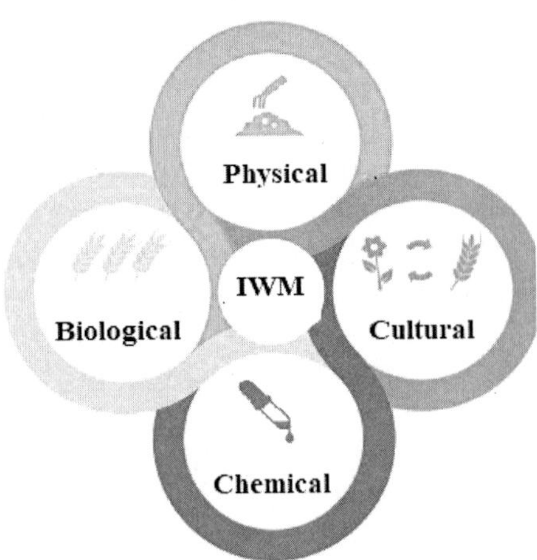

Fig. 5.3: Integration four sciences/means of managing weeds

Integrated weed management is therefore the process of using all four sciences, in a whole-of-system, or integrated fashion to achieve the best weed management. This does not mean that all four sciences will be brought to bear on all weed problems, rather, the optimal tools will be chosen for each specific weed issue. This is where the "many little hammers" (Liebman and Gallandt, 1997) or "many tools in the toolbox" and "Swiss army knife" metaphors originate. The key is that, unlike herbicides that can be used in isolation from other farm activities, integrated weed management requires a range of tools to be used in an integrated fashion to achieve overall weed management, for example, integrating rotations, with tillage, and hoeing. In organics, and increasingly in mainstream farming, it is also, therefore, essential to integrate weed management with all other farm activities, such as other crops, tillage systems, soil management, pest and disease management, etc., to create "integrated farm management."

5.3.2 Way of Weed Control Management

5.3.2.1 Physical Management

Physical management is the removal of weeds by physical or mechanical means, such as mowing, grazing, mulching, tilling, burning or by hand.

5.3.2.1.1 Burning: Burning removes the above-soil body of the weeds killing most of the plants. If carried out before seed is set it can prevent the further spread of weeds. Burning can be undertaken over a wide area with minimal human input. As with tilling, burning exposes the soil surface to erosion. If burning is used as a control method, caution should be exercised to minimise the risk of harm to the environment and to those undertaking the activity.

5.3.2.1.2 Hay making, mowing and grazing: Hay making, mowing and grazing before weeds produce seeds restrict the amount of weed seed in an area and reduce the spread of weeds.

5.3.2.1.3 Hand removal: Removal by hand, including hoeing, is a good method for selective removal of weeds without disturbing the surrounding desirable vegetation. It is very labour intensive and is often only used in small areas, such as gardens or in larger areas during bush regeneration.

5.3.2.1.4 Mulching: Mulching, by covering the ground with a layer of organic material, suppresses or kills weeds by providing a barrier between the weeds and sunlight. Mulching has an added advantage in that it improves the condition and moisture level in the soil. Planting competitive and desirable plants that provide a dense cover over the weeds suppresses weed growth in a similar way to mulching.

5.3.2.1.5 Tilling: Tilling, the ploughing or cultivation method that turns over the soil, buries the weed beneath the soil. This provides a barrier to the sun, therefore killing the weeds. Tilling is a form of physical control that can be easily undertaken over a wide area, using agricultural machinery. This method is useful for making soil ready for planting new crops, but it can lead to damage in soil structure and exposes the soil to erosion and further invasion by weeds.

5.3.2.1.6 Hoeing: Hoe has been the most appropriate and widely used weeding tool for centuries. It is however, still a very useful implement to obtain results effectively and cheaply. It supplements the cultivator in row crops. Hoeing is particularly more effective on annuals and biennials as weed growth can be completely destroyed. In case of perennials, it destroyed the top growth with little effect on underground plant parts resulting in re-growth.

5.3.2.1.7 Digging: Digging is very useful in the case of perennial weeds to remove the underground propagating parts of weeds from the deeper layer of the soil.

5.3.2.1.8 Flooding: Flooding is successful against weed species sensitive to longer periods of submergence in water. Flooding kills plants by reducing oxygen availability for plant growth. The success of flooding depends upon complete submergence of weeds for longer periods.

5.3.2.1 Cultural Management

Several cultural practices like tillage, planting, fertiliser application, irrigation etc., are employed for creating favourable condition for the crop. These practices if used properly, help in controlling weeds. Cultural methods, alone cannot control weeds, but help in reducing weed population. They should, therefore, be used in combination with other methods. In cultural methods, tillage, fertiliser application and irrigation are important. In addition, aspects like selection of variety, time of sowing, cropping system, cleanliness of the farm etc., are also useful in controlling weeds.

5.3.2.2.1 Field Preparation: The field has to be kept weed free. Flowering of weeds should not be allowed. This helps in prevention of buildup of weed seed population.

5.3.2.2.2 Highly Qualitative and Yielding Cultivars: Use high quality (large and plump) seeds, as they are more likely to produce vigorous and competitive plants. High yielding cultivars are less competitive against weeds than traditional cultivars. For rainfed areas, heavy tillering varieties of medium stature may be better suited than semi-dwarf varieties.

5.3.2.2.3 Timely nutrient application: Nutrient application should be timed to prevent weed proliferation and yet to obtain maximum benefit from the applied nutrient.

5.3.2.2.4 Maintenance of optimum plant population: Lack of adequate plant population is prone to heavy weed infestation, which becomes, difficult to control later. Therefore practices like selection of proper seed, right method of sowing, adequate seed rate' protection of seed from soil borne pests and diseases etc. are very important to obtain proper and uniform crop stand capable of offering competition to the weeds. Use shallow seeding techniques, where possible, to allow the desired species to grow above the soil surface more quickly.

5.3.2.2.5 Crop rotation: The possibility of a certain weed species or group of species occurring is greater if the same crop is grown year after year. In many instances, crop rotation can eliminate at least reduce difficult weed problems. The obnoxious weeds like *Cyperus rotundus* can be controlled effectively by including low land rice in crop rotation. Some practices that make it hard for weeds to adapt and therefore reduce their spread and vigour include i.e. (a) Rotate species with different seasonal and growing cycles, and (b) Rotate herbicides with different modes of action to help delay the development of herbicide resistance.

5.3.2.2.6 Intercropping: Inter cropping suppresses weeds better than sole cropping and thus provides an opportunity to utilize crops themselves as tools of weed management. Many short duration pulses viz., green gram and soybean effectively smother weeds without causing reduction in the yield of main crop.

5.3.2.2.7 Mulching: Mulch is a protective covering of material maintained on soil surface. Mulching has smothering effect on weed control by excluding light from the photosynthetic portions of a plant and thus inhibiting the top growth. It is very effective against annual weeds and some perennial weeds like *Cynodon dactylon*. Mulching is done by dry or green crop residues, plastic sheet or polythene film. To be effective the mulch should be thick enough to prevent light transmission and eliminate photosynthesis.

5.3.2.2.8 Blind Tillage: The tillage of the soil after sowing a crop before the crop plants emerge is known as blind tillage. It is extensively employed to minimise weed intensity in drill sowing crops where emergence of crop seedling is hindered by soil crust formed on receipt of rain or irrigation immediately after sowing.

5.3.2.2.9 Summer Tillage: The practice of summer tillage or off-season tillage is one of the effective cultural methods to check the growth of perennial weed population in crop cultivation. Initial tillage before cropping should encourage clod formation. These clods, which have the weed propagules, upon drying desiccate the same. Subsequent tillage operations should break the clods into small units to further expose the shriveled weeds to the hot sun.

5.3.2.3 Biological Management

The biological control approach makes use of the invasive plant's naturally occurring enemies, to help reduce its impact. It aims to reunite weeds with their natural enemies and achieve sustainable weed control. These natural enemies of weeds are often referred to as biological control agents. Biological weed control using insects, pathogens, fish and snails (bio agents) appears to be ideal for reducing the inputs of herbicides (Table 5.1). Although herbicides are effective for weed control, there has been increasing concern about their safety for food products, their adverse effect on environment and widespread weed resistance to herbicides (Table 5.2)

Table 5.1: List of Weed and Biocontrol agent/s

Weed	Biocontrol agent/s
Alternanthera philozeroid	Cassida sp.
Cyperus rotundus	*Bactra minima* (insect) and *Athespacuta cyperi* (weevil)
Eichhornia crossipes	*Alternaria eichhornia* (pathogen) and *Neochetina bruchi* (insect)
Parthenium histerophrous	*Zygograma bicolorata* and *Smicronyx lutulentus*
Salvania molesta	*Paulinia acuiminata* (insect) and *Myrothecium rovidium* (fungus)

Table 5.2: List of bio-herbicides with Targeted weed, Bioagent and Crop/s

Bioherbicide	Target weed	Bioagent	Crop/s
Bactra verutana	*Cyperus rotundus*	Shoot boring moth	Rice and wheat
Biomal	*Malva pusilla*	Colletitrichum gloeosporioides	Row crops
Biopolaris	*Sorghum halepense*	Bipolaris sorghicola	Rice and wheat
BioChon	*Prunus serotina*	Chondrostereum purpureum	Forests
Collego	*Aschyynomene virginica*	Colletotrichum gloeosporioides	Rice
DeVine	*Morrenia odorata*	Phytophthora palmovora	Citrus groves
Emmalocera sp	*Echinochloa* sp	Stem boring moth	Rice and wheat
Gastrophysa	*Rumex sp*	Beetle	Rice and wheat
Tripose	*Echinochloa* sp	Shrimp	Rice and wheat
Uromyces rumicis	*Rumex* sp	Plant pathogen	Rice and wheat

Source: Rana, 2016

5.3.2.4 Chemical Management

Although the use of chemicals is not always essential, herbicides can be an important and effective component of any weed control programme. In some situations herbicides offer the only practical, cost-effective and selective method of managing certain weeds. Because herbicides reduce the need for cultivation, they can prevent soil erosion and water loss, and are widely used in conservation farming. In some cases, a weed is only susceptible to one specific herbicide and it is important to use the correct product and application rate for control of that particular weed. Hand weeding is ineffective against perennial weeds due to their regenerative capability. Several research publications (Singh *et al.,* 1999; Singh *et al.,* 2001a; Rameshwar *et al.,* 2002; Sardana *et al.,* 2006; Rao and Nagamani, 2007; Nagar *et al.,* 2009) have proved that integration of herbicides with hand weeding is the most effective and economical method of weed management. Rao and Nagamani (2010) summarized economical weed management strategies for a few major crops of India (Table 5.3).

Table 5.3: Most economical IWM methods for managing weeds in certain crops of India

Crop	IWM	Reference/s
Asgandh (Withania somnifera Dunal)	PE of isoproturon at 0.50 kg/ha and glyphosate at 1.0 kg/ha fb HW 45 DAS	Kulmi and Tiwari (2005)
Blackgram	(a) Pendimethalin at 0.75 kg/ha fb HW 45 DAS (b) Pendimethalin at 0.50 kg/ha fb HW 60 DAS (c) Trifluralin (PE) at 0.50 kg/ha fb HW	(a) Kumar *et al.*, (2006) (b) Rathi *et al.*, (2004) (c) Sardana *et al.*, (2006)
Coriander	Pendimethalin (PE) at 1.0 kg/ha fb HW 45 DAS	Nagar *et al.*, (2009)
Cowpea	Pendimethalin 0.75 kg/ha fb HW 35 DAS	Mathew *et al.*, (1995)
Garlic	PE of oxyfluorfen (0.15 kg/ha) or pendimethalin (1.0 kg/ha) fb HW 40 DAS	Porwal (1995)
Groundnut	Pendimethalin or alachlor 1 kg/hafb HW 30 DAS	Itnal *et al.*, (1993)
Indian mustard	(a) Pendimethalin (PE) at 0.50 kg/ha or fluchloralin at 0.50 kg/ha each fb HW 30 DAS (b) Fluchloralin at 0.75 kg/ha fb HW 25 DAS	(a) Singh *et al.*, (1999) (b) Singh (2006)
Onion	(a) Pendimethalin at 1.5 kg/ha fb HW 60 DAT (b) Oxyfluorfen applied at 0.25 kg/ha fb HW 40 DAT (c) Oxyfluorfen at 0.15 kg/ha fb HW 35 DAT (d) Fluchloralin or pendimethalin at 0.9 kg/hafb HW 40 DAT	(a) Rameshwar *et al.* (2002) (b) Nandal and Singh (2002) (c) Kolhe (2001) (d) Sukhadia *et al.* (2002)
Okra	Stale seed bed with glyphosate application integrated with eucalyptus mulching	Ameena *et al.*, (2006)
Opium poppy (Papaver somniferum L.)	Isoproturon at 375 g/ha or 500 g/ha PE fb HW 30 DAS	Kulmi and Tiwari (2004)
Dwarf pea	Sowing at 20 cm apart with two HW fb pendimethalin at 1 kg/ha	Tewari *et al.,* (2003)
Pigeonpea/ Groundnut intercrop	Pendimethalin (1.0 kg/ha) or fluchloralin (1.0 kg/ha) each fb two HW 30 and 42 DAS	Vijaykumar *et al.*, (1995)

Crop	IWM	Reference/s
Pigeonpea/pearl millet intercrop	Pendimethalin at 1.50 kg/ha + HW 40 DAS	Shinde *et al.*, (2003)
Rice-transplanted rice	(a) Application of butachlor 1.0 kg/ha, anilofos 0.4 kg/ha along with closer planting (b) Anilophos 0.6 kg/ha 7 DAT+HW 27 DAT	(a) Gogoi *et al.*, (2001) (b) Singh and Kumar (1999)
Rice-dry-seeded rice	Butachlor at 1.0 kg/ha fb one hand weeding at 30 DAS by local tool 'Kutla'	Singh and Singh (2001)
Sesame	60 kg N/ha + fluchloralin at pre-planting@ 1.0 kg/ha fb HW 21 DAS	Singh *et al.*, (2001c)
Soybean	(a) Butachlor (Pre-em) 1.5 kg/ha fb HW 30 DAS (b) Rows spacing of 22.5 cm with alachlor at 1 kg/ha	(a) Chandrakar and Urkurkar (1993) (b) Shekara and Nanjappa (1993)
Sugarcane	Metribuzin or atrazine at 1 kg/ha + trash mulching (3.5 t/ha) in between cane rows at 60 DAP	Singh *et al.*, (2001b)
Wheat	Pendimethalin at 0.75 kg/ha + HW 30 DAS	Singh and Singh (2004)

DAS–Days after seeding, DAT–Days after transplanting, DAP–Days after planting, HW–Hand weeding, fb–followed by, PE–Pre-emergence. (The crops are mentioned alphabetically, not according to their economic importance) Herbicide rate: kg ha^{-1} or g ha^{-1}.

5.4 Integrated Plant Nutrient Management

INM refers to the maintenance of soil fertility and of plant nutrient supply at an optimum level for sustaining the desired productivity through optimization of the benefits from all possible sources of organic, inorganic and biological components in an integrated manner. Soil is a fundamental requirement for crop production as it provides plants with anchorage, water and nutrients. A certain supply of mineral and organic nutrient sources is present in soils, but these often have to be supplemented with external applications, or fertilisers, for better plant growth. Fertilisers enhance soil fertility and are applied to promote plant growth, improve crop yields and support agricultural intensification. Optimal and balanced use of nutrient inputs from mineral fertilisers will be of fundamental importance to meet growing global demand for food (International Food Policy Research Institute, 1995). Mineral fertiliser use has increased almost fivefold since 1960 and has significantly supported global population growth. Efficient use of all nutrient sources including organic sources, recyclable wastes, mineral fertilisers and biofertilisers should therefore be promoted through Integrated Nutrient Management (Roy *et al.,* 2006), in wheat (Mishra, 2012; Karkalia *et al,* 2017) in rice (Mondal *et al.*, 2019). INM relies on a number of factors, including appropriate nutrient application and conservation and the transfer of knowledge about INM practices to farmers and researchers. Boosting plant nutrients can be achieved by a range of practices covered in this guide such as terracing, alley cropping, conservation tillage, intercropping, and crop rotation. Given that these technologies are covered elsewhere in this guidebook, this section will focus on INM as it relates to appropriate fertiliser use. In addition to the standard selection and application of fertilisers, INM practices include new techniques such as deep placement of fertilisers and the use of inhibitors or urea coatings (use of urea coating agent helps to restart the activity and growth of the bacteria responsible for denitrification) that have been developed to improve nutrient uptake. The major concept of IPNM is (a) Regulated nutrient supply for optimum crop growth and higher productivity (b) Improvement and maintenance of soil fertility, and (c) Zero adverse impact on agro ecosystem quality by balanced fertilization of organic manures, inorganic fertilizers and bio- inoculant. The field level management practices considered under the heading of IPNM would include the use of farmyard manures, natural and mineral fertilizers, soil amendments, crop residues and farm wastes, agroforestry and tillage practices, green manures, cover crops, legumes, intercropping, crop rotations, fallows, irrigation, drainage, plus a variety of other agronomic, vegetative and structural measures designed to conserve both water and soil. The underlying principles on how best to manage soils, nutrients, water, crops and vegetation to improve and sustain soil fertility and land productivity and their processes are derived from the essential soil functions necessary for plant growth. The following are fundamental to the approach outlined for IPNM:

- Loss of soil productivity is much more important than the loss of soil itself, thus land degradation should be prevented before it arises, instead of attempting to cure it afterwards i.e. the focus for IPNM should be on sustaining the productive potential of the soil resource.
- Soil and plant nutrient management cannot be dealt with in isolation but should be promoted as an integral part of a productive farming system.
- Under rainfed dry land farming conditions soil moisture availability is the primary limiting factor on crop yields, not soil nutrients as such, hence IPNM requires the adoption of improved rainwater management practices (conservation tillage, tied ridging etc), so as to increase the effectiveness of the seasonal rainfall.

- With declining soil organic matter levels following cultivation, the adoption of improved organic matter management practices are a prerequisite for restoring and maintaining soil productivity (improved soil nutrient levels, soil moisture retention, soil structure and resistance to erosion).

Components of Integrated Nutrient Management

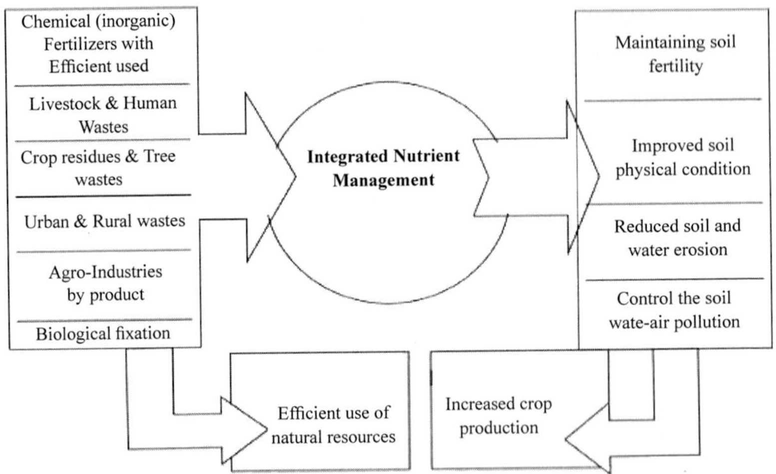

Fig 5.4: Resources of INM and their role in Soil Productivity

5.4.1 Integrated and synergistic approach for IPNM

It is only after they have made improvements in the biological, physical and hydrological properties of their soils that farmers can expect to get the full benefits from the supply of additional plant nutrients, in the form of inorganic fertilizer, to their crops. According to FAO, at the farm field level IPNM therefore calls for an integrated and synergistic approach which involves:

Better plant management, especially (i) improved crop establishment at the beginning of the rains, so as to increase protective ground cover thereby reducing splash erosion, enhancing infiltration and biological activity; and (ii) timely weeding to reduce crop yield losses from competition for nutrients and soil moisture.

- Combinations of complementary crop, livestock and land husbandry practices which maximize additions of organic materials and recycle farm wastes, so as to maintain and enhance soil organic matter levels (ideally at levels of at least 50-75% of those under natural vegetation).
- Combinations of crop, livestock and land husbandry practices that reduce rainfall impact, improve surface infiltration, and reduce the velocity of surface runoff thereby ensuring any soil loss is below the 'tolerable' level for the soil type.
- Conservation tillage, crop rotation, agroforestry and restorative fallow practices that maintain and enhance the soils physical properties through maintaining an open topsoil structure, and breaking any subsoil compacted layer (hoe/plough pan) thereby encouraging root development and rainfall infiltration (e.g. use of ox drawn chisel ploughs, double dug beds, pasture leys, interplanting of deep rooted perennial crops/trees and shrubs).
- Land management practices that ensure soil moisture conditions are favourable for the proposed land use (e.g. moisture harvesting/ conservation in low rainfall areas, drainage in high rainfall areas).
- Matching the land use requirements of individual agricultural enterprises with the land qualities present in the areas where they are undertaken - i.e. the biological, chemical and physical properties of the soil, and the local climatic conditions (temperature, rainfall etc).
- Seeking to improve yields by identifying and overcoming the most limiting factors in order of their diminishing influence on yield.
- The replenishment of soil nutrients lost by leaching and/or removed in harvested products through an integrated plant nutrition management approach that optimizes the benefits from all possible on- and off-farm sources of plant nutrients (e.g. organic manures, crop residues, rhizobial N-fixation, P and other nutrient uptake through root mycorrhizal fungi infestation, transfer of nutrients released by weathering in the deeper soil layers to the surface via tree roots and leaf litter, rock phosphate, inorganic fertilizer etc.).

Table 5.4: INM in some commercial Vegetable crops

Crop	INM Option
Brinjal (*Solanum melongena*)	Application of 50 kg nitrogen through urea along with 50 kg nitrogen through poultry manure per hectare help in increasing yield of brinjals and sustained soil health
Okra (*Abelmoschus esculentus*)	1. In Assam application of N, P, K @ 37.5:37.5:37.5 kg per hectare along with vermicompost @ 1 t per hectare mixed with microbial consortia @ 3.5 kg per hectare increased fruit yield and also sustaining soil health 2. Combination of N, P, K @ 50:50:50 kg per hectare along with FYM @ 10 t per hectare found beneficial in increasing yield.
Tomato (*Lycopersicon esculentum*)	Incorporation of FYM @ 40 t per hectare with half dose of NPK (75:30:30) kg per hectare substituted 50% of recommended dose of fertilizer.
Chilli (*Capsicum annum*)	At Dapoli (Maharashtra) FYM @ 10 t per hectare + 50% recommended dose of fertilizers proved beneficial.
Potato (*Solanum tuberosum*)	Application of 75% recommended dose of fertilizer along with incorporation of 20 t per hectare increased tuber yield of potato in clay soils of Palampur.
Onion (*Allium cepa*)	At Wadura of Jammu and Kashmir application of 52.5 kg nitrogen per hectare with use of azotobacter biofertilizer recorded higher onion yield.

Source: Kumar and Ramjan (2019)

Table 5.5: Plant Protection Products (PPPs) in Organic Farming

Name of product	Purpose and specifications of use
Azadirachtin	Azadirachtin from the neem tree (Azadirachta indica)
Beeswax	Used as protectant for treatment of cuts and wounds after pruning or in grafting
Plant oils	Used for control of small-bodied insects such as thrips, aphids, and whiteflies
Laminarin (from *Laminaria digitata*) or kelp or brown algae seaweed	A polysaccharide from the group of the glucans, used to protect plants against fungi and bacteria. Kelp should be grown according to the organic standards
Pheromones	Used only in traps and dispensers
Pyrethrins from the leaves of *Chrysanthemum cinerariaefolium*	Used as insecticide
Pyrethroids (only deltamethrin or lambdacyhalothrin)	Used only in traps with attractants or pheromones
Quassia from the plant *Quassia amara*	Only insecticide and repellent
Microorganisms, e.g., *Bacillus* thuringiensis, *Beauveria bassiana*, and *Metarhizium anisopliae*	Origin should not be GMOs

Name of product	Purpose and specifications of use
Spinosad from the soil bacterium *Saccharopolyspora spinosa*	Used as insecticide
Ethylene	Insecticidal fumigant against fruit flies
Parafin oil	Used as insecticide against small-bodied insects
Fatty acids (soft soaps)	Insecticide against mite, thrips, and aphids
Lime sulfur (mixture of calcium hydroxide and sulfur)	Used as fungicide
Kieselgur (diatomaceous earth) from the hard shelled diatom protist (chrysophytes)	Used as mechanical insecticide
Naturally occurring aluminum silicate (kaolin)	As insect repellent against a wide range of insects at a rate of 50 kg/ha
Calcium hydroxide	Used as fungicide
Sodium hypochlorite (bleach or as javel water). It is a disinfectant with numerous uses, and its effect is due to the chlorine	Used in seed treatment as viricide and bactericide
Sulfur	Used as broad-spectrum inorganic contact fungicide and acaricide
Copper compounds such as: copper hydroxide, copper oxychloride, copper oxide, tribasic copper sulfate, and Bordeaux mixture (copper sulfate and calcium hydroxide)	Used as fungicide and bactericide maximum of 6 kg/ha copper annually
Sheep fat (obtained from fatty sheep tissues by heat extraction and mixed with water to obtain an oily water emulsion)	A triglyceride consisting predominantly of glycerine esters of palmitic acid, stearic acid, and oleic acid. A repellent by smell against vertebrate pests such as deer and other game animals. It should not be applied to the edible parts of the crop
Quartz sand	Used as repellent against vertebrate pests

Source: Commission Regulation (EC) No 889/2008 of 5 September 2008

5.5 Integrated Farm Management (IFM)

Integrated farming is a whole farm management system which aims to deliver more sustainable agriculture. It is a dynamic approach which can be applied as the best of modern technology and traditional methods to any farming system around the world. Thus, it involves attention to detail and continuous improvement in all areas of a farming business through informed management processes. Integrated Farm Management (IFM) is a site-specific farm trade approach, and made up of nine sections. Each section is interrelated and an understanding of how they work together is essential for the effective implementation of IFM.

Fig 5.5: Integrated Farm Management System

5.5.1 Advantages of IFM

- Adoption of new technology
- Avoiding deforestation
- Balanced food
- Environment safety
- Income round the year
- Increased productivity and profitability
- Input-output efficiency
- Recycling of resources
- Solving energy (fuel and fodder crises)
- Sustainability

5.5.2 Factors affecting an integrated farming

- Availability of the resources, land and labor
- Economics of proposed integrated farming system
- Management skill of farmer.
- Present level of utilization of resources
- Soil and climatic conditions of the selected area

5.6 Manures

Manures are plant and animal wastes that are used as sources of plant nutrients. They release nutrients after their decomposition. Manures can be grouped into bulky organic manures and concentrated organic manures based on concentration of the nutrients (TNAU Portal; Rana, 2016)

5.6.1 Bulky Organic Manures: Bulky organic manures contain small percentage of nutrients and they are applied in huge quantities. Farmyard manure (FYM), compost and green manure are the most important and widely used bulky organic manures. Use of bulky organic manures have several advantages:-

a) They supply plant nutrients including micronutrients,

b) They improve soil physical properties like structure, water holding capacity etc.,

c) They increase the availability of nutrients, Carbon dioxide released during decomposition acts as a CO_2 fertilizer, and

d) Plant parasitic nematodes and fungi are controlled to some extent by altering the balance of microorganisms in the soil.

5.6.2 Farmyard Manure: This is the traditional manure and is mostly readily available to the farmers. FYM refers to the decomposed mixture of dung and urine of farm animals along with litter and left over material from roughages or fodder fed to the cattle. Cattle manure is slow acting, bulky organic and however is a low analysis fertilizer, obtained from dung and urine of farm animals mixed with litter and other miscellaneous farm wastes. On an average well decomposed farmyard manure contains 0.5% N, 0.2% P_2O_5 and 0.5% K_2O. The present method of preparing farmyard manure by the farmers is defective. Urine, which is wasted, contains one % nitrogen and 1.35% potassium. Nitrogen present in urine is mostly in the form of urea which is subjected to volatilization losses. Even during storage, nutrients are lost due to leaching and volatilization. However, it is practically impossible to avoid losses altogether, but can be reduced by following improved method of preparation of farmyard manure.

5.6.2.1 Quality and Composition of FYM

The quality of manure and chemical composition in particular is highly variable depending upon the kind of animal, age and condition of the individual animal, quality and quantity of feed consumed, kind of litter used, collection and storage of manure.

5.6.2.1.1 Parameters for Animals: Growing animals, milch cattle, pregnant or carrying cattle utilize much of the ingredients in the feeds for building up their growing bodies, milk production and for the development of the embryo [calf]. Old or adult animals kept on light work or no work, utilize little from feeds and as such, most of nitrogen is voided through urine and dung. Eventually, the adult old cattle provide better manure. The quality of manure depends on the class of manure viz., cattle, horse manure. Within the same class, quality varies according to the kind of animal, such as milch cattle, dry cattle, work cattle, breeding bulls etc. By and large the dung and urine from animals, which assimilates less (little) for their maintenance and production, will provide better quality manure.

5.6.2.1.2 Litter and Feed: The quality of manure depends to a considerable extent on the nature of litter used as well as quality and quantity of consumed feed. Remnants of leguminous hays (*Bhusa*) give richer manure than usual straws. Nutritious and protein rich feeds like oil cakes enriches the nitrogen content to the resulting manure than the bulky feeds like straw and green grass. Animals fed on concentrated feeds yield better quality manure.

5.6.2.1.3 Manure Collection: The method adopted for collection of dung, urine and litter primarily decide the quality of manure as the loss of nutrients particularly nitrogen occurs from the time urine and dung are voided by cattle. The quality of manure depends upon the methods of collection *viz*., Byre, Lose box and Dry earth systems

- **Byre system:** Cattle are stalled in a shed with a non-absorbent floor provided with necessary slope towards the urine drains. The urine that flows into the drains is collected into a covered tank. From where it is periodically removed and sprinkled on the manure stored in a covered pit .The urine which is an important component of FYM can be properly stored (conserved) in this system. The perfect cleanliness and hygienic conditions of the stalls as well as cattle can be maintained in this system.
- **Dry Earth System:** The floor of the cattle shed is well rammed and compacted. Layers of fine sand, red earth of loamy soil are spread as an absorbent for urine. The wet portions are properly covered with dry layers or any of the above materials and once a week the surface layer is removed and dumped in the manure pit. Available saw dust, paddy husk, groundnut shell, paddy winnowed dust would serve the purpose very well compared to the earth absorbents. This system is popular and extensively adopted in rural parts being cheap, convenient and practicable under the existing rural conditions in India.

5.6.2.2 Manure Storage: Method of storage of manure influences the quality of manure to a large extent. During storage the manure undergoes fermentative changes, decomposition which leads to losing its original structure and shape. There are three methods of storage i.e pit method, heap method and covered pit method.

5.6.2.2.1 Pit Method: According to this method, the manure is stored in a pit with non-absorbent bottom and sides (below ground level). The pit is provided with a bund at the rim of the pit to prevent the surface run-off of waters during rainy season. The dimensions of the pit can be variable depending on the quantity of dung, urine and litter produced on the farm per day. The losses also occur in this method due to exposure to sun and rain, but it is relatively a better method than the heap method.

5.6.2.2.2 Heap Method: This is the most common method adopted in Indian villages. According to this method, manure is heaped on the ground preferably under the shade of a tree (above the ground level). Ideal procedure is to dump the dung first and to cover it with litter soaked urine. This is further covered with a layer of litter/ash / earth to prevent the loss of moisture and to avoid direct exposure to sun. It is also desirable to put up a small bund around the base of the heap to protect against surface run-off washing out the manurial ingredients. It is beneficial to cover the exposed portion of the heap with Palmyra leaves or any other available material. The maximum losses of nutrients occur in this method of storage, resulting in poor quality manure. Direct exposure to the vagaries of climate such as sunshine and rainfall causes looseness and dryness of manure, which hasten the losses of nutrients and rapid oxidation of organic matter.

5.6.2.2.3 Covered Pit Method: According to this method, the bottom and sides of the pit are made non-absorbent by granite stone lining. The pit is also provided with a bund of 1½ feet height to prevent surface flow of water (rain water) and a suitable cover by way of roofing with locally available materials like Palmyra or phoenix leaves etc., organic matter and nutrient losses can be effectively controlled in this method of storage in order to obtain better quality manure (FYM: 0.68% N, 0.5% P, 1% K).

5.6.3 Goat and Sheep Manure

The dropping of goats and sheep contain higher nutrients than farmyard manure and compost. On an average, the manure contains 3% N, 1% P_2O_5 and 2% K_2O. It is applied to the field in two ways i.e. (a) Sweeping of sheep or goat sheds are placed in pits for decomposition and it is applied later to the field. The nutrients present in the urine are wasted in this method. (b) Sheep penning, wherein sheep and goats are allowed to stay overnight in the field and urine and fecal matter is added to the soil which is incorporated to a shallow depth by running blade harrow or cultivator.

5.6.4 Poultry Manure

The excreta of birds ferments very quickly. If left exposed, 50 % of its nitrogen is lost within 30 days. Poultry manure contains higher nitrogen and phosphorus compared to other bulky organic manures. The average nutrient content is 3.03 % N, 2.63 % P_2O_5 and 1.4 % K_2O.

5.6.5 Compost

A mass of rotted organic matter made from waste is called compost. The compost made from farm waste like sugarcane trash, paddy straw, weeds and other plants and other waste is called farm compost. The average nutrient content of farm compost is 0.5 % N, 0.15 % P_2O_5 and 0.5 % K_2O. The nutrient value of farm compost can be increased by application of superphosphate or rock phosphate at 10 to 15 kg/t of raw material at the initial stage of filling the compost pit. The compost made from town refuses like street sweepings and dustbin refuse is called town compost. It contains 1.4 % N, 1.00 % P_2O_5 and 1.4 % K_2O. Farm compost is made by placing farm wastes in trenches of suitable size, say, 4.5 to 5.0m long, 1.5 to 2.0m wide and 1.0 to 2.0m deep. Farm waste is placed in the trenches layer by layer. Each layer is well moistened by sprinkling cow-dung slurry or water. Trenches are filled up to a height of 0.5m above the ground. The compost is ready for application within 5 to 6 months. Compost prepared by traditional method is usually low in nutrients and there is need to improve its quality. Enrichment of compost using low cost N fixing and phosphate solubilizing microbes is one of the possible ways of improving nutrient status of the product. It could be achieved by introducing microbial inoculants, which are more efficient than the native strains associated with substrate materials. Both the nitrogen fixing and phosphate solubilizing microbes are more exacting in their physiological and ecological requirements and it is difficult to meet these requirements under natural conditions. The only alternative is to enhance their inoculums potential in the composting mass. Studies conducted at IARI, New Delhi, showed that inoculation with *Azotobacter/Azospirillum* and phosphate solubilising culture in the presence of 1% rock phosphate is a beneficial input to obtain good quality compost rich in N (1.8%). The humus content was also higher in materials treated with microbial inoculants.

The following basic rules are important for the production of good quality compost:

- The purpose of composting is to convert organic matter into growth promoting substances, for sustained soil improvement and crop production.
- Certain additives accelerate the conversion and improve the final product. The materials such as lime, earth, gypsum, rock phosphate act as effective additives. The addition of nitrogen (0.1 to 1%) is important in case of large C:N ratio of the composting material. Addition of lime (0.3 to 0.5 %), if sufficient lime is not present. The preparation of compost takes 2-3 months. The composition of compost varies with in wide limits.
- Soil microorganisms constitute sufficiently to the decomposition of organic matter through their continuous activities. The majority of these soil animals provide optimal conditions in their digestive track for their synthesis of valuable permanent humus and stable soil crumbs (typical compost earthworm is *Eisenia foetida.*).
- The organic matter is partially decomposed and converted by microbes. These microbes require proper growth conditions, for their activity (i.e Moisture Content: 50% and 50% aeration of total pore space of the composting material). This is achieved through stacking and occasional turning over. Microbes also need sufficient nitrogen for synthesizing their body cells (the optimum C:N ratio of the composting material is 20:1 to 30:1).

5.6.5.1 Stages of Composting

There is a huge difference between backyard manure and municipal composter. Municipal composters handle large batches of organic materials all at once and known as batch composter, while backyard composters continuously produce a small amount of organic material every day, and known as continuous composters. During the time of composting of organic material in a batch, four stages are apparent. Although the same phases occur during continuous composting, while they are not as apparent as in a batch, and, in fact, they may be occurring concurrently rather than sequentially. The four phases of composting is given here as:

a. Mesophilic Phase

b. Thermophilic Phase

c. Cooling Phase, and

d. Curing Phase.

a) Mesophilic Phase

This is the first stage of the composting process. In this phase, compost bacteria combine carbon with oxygen to produce carbon dioxide and energy. Some of the energy is used by the microorganisms for reproduction and growth, the rest is given off as heat. When a pile of organic refuse begins to undergo the composting process, mesophilic bacteria proliferate, raising the temperature of the composting mass up to 44°C (111°F).

b) Thermophilic Phase

Here, mesophilic bacteria (*E. coli* and other bacteria from the human intestinal tract) soon become increasingly inhibited by the temperature, as the thermophilic bacteria take over in the transition range of 44°C to 52°C (111°F to 125.6°F), and begins the second stage of the process, when thermophilic microorganisms are very active and produce a lot of heat. This stage can then continue up to about 70°C (158°F), although such high temperatures are neither common nor desirable in backyard compost. This heating stage takes place rather quickly and may last only a few days, weeks, or months. It tends to remain localized in the upper portion of a backyard compost bin where the fresh material is being added, whereas in batch compost, the entire composting mass may be thermophilic all at once. After the thermophilic heating period, the humanure will appear to have been digested, but the coarser organic material will not. This is when the third stage of composting, the cooling phase takes place.

c) Cooling Phase

During this phase, the microorganisms that were chased away by the thermophiles migrate back into the compost and get back to work digesting the more resistant organic materials. Fungi and macroorganisms such as earthworms and sowbugs that break the coarser elements down into humus also move back in. After completion of the thermophilic staged, only the readily available nutrients in the organic material have been digested. There's still a lot of food in the pile, and a lot of work to be done by the creatures in the compost. It takes many months to break down some of the more resistant organic material in compost such as “lignin” which comes from wood materials. Like humans, trees have evolved with a skin that is resistant to bacterial attack, and in a compost pile those lignins resist breakdown by thermophiles. However, other organisms, such as fungi, can break down lignin, given enough time; since they don't like the heat of thermophilic compost, they simply wait for things to cool down before beginning their job.

d) Curing Phase

The final stage of the composting process is called the curing, aging, or maturing stage, and it is a long and important one. Commercial composting professionals often want to make their compost as quickly as possible, usually sacrificing the compost's curing time. One municipal compost operator remarked that if he could shorten his compost time to four months, he could make three batches of compost a year instead of only the two he was then making, thereby increasing his output by 50%. Municipal composters see truckloads of compost coming in to their facilities daily, and they want to make sure they don't get inundated with organic material waiting to be composted.

Therefore, they feel a need to move their material through the composting process as quickly as possible to make room for the new stuff coming in. Household composters don't have that problem, although there seem to be plenty of backyard composters who are obsessed with making compost as quickly as possible. However, the curing, aging, or maturing of the compost is a critically important stage of the compost-making process. And, as in wine-making, an important element to figure into the equation is patience.

A long curing period (e.g., a year after the thermophilic stage) adds a safety net for pathogen destruction. Many human pathogens only have a limited period of viability in the soil, and the longer they are subjected to the microbiological competition of the compost pile, the more likely they will die a swift death. Immature compost can be harmful to plants. Uncured compost can produce phytotoxins (substances toxic to plants), can rob the soil of oxygen and nitrogen, and can contain high levels of organic acids. So relax, sit back, put your feet up, and let your compost reach full maturity *before* you even think about using it.

5.6.6 Green Manure

Green un-decomposed plant material used as manure is called green manure. It is obtained in two ways i.e (a) by growing green manure crops or (b) by collecting green leaf (along with twigs) from plants grown in wastelands, field bunds and forest. Green manuring is growing in the field plants usually belonging to leguminous family and incorporating into the soil after sufficient growth. The plants that are grown for green manure are known as green manure crops.

Green manures are crops grown for the express purpose of plowing them in, thus increasing fertility through the incorporation of nutrients and organic matter into the soil. Leguminous plants such as clover are often used for this, as they fix nitrogen using *Rhizobia* bacteria in specialized nodes in the root structure. Other types of plant matter used as manure include the contents of the rumens of slaughtered ruminants, spent grain (left over from brewing beer) and seaweed. This phosphorus is available to succeeding crop after mineralization of the incorporated green manure crop. Application to the field, green leaves and twigs of trees, shrubs and herbs collected from elsewhere is known as green-leaf manuring. Forest tree leaves are the main sources for green-leaf manure. Plants growing in wastelands, field bunds etc., are another source of Greenleaf manure. The important plant species useful for green-leaf manure are Neem, Mahua, Dhaincha, Pillipesara, Wild indigo, Glyricidia, Karanji (*Pongamia glabra*) Calotropis, Avise (*Sesbania grandiflora*), Subabul and other shrubs. Growing deep rooted green-manure crops and their incorporation facilitates in bringing nutrients to the top layer from deeper layers. Nutrient availability increases due to production of carbon dioxide and organic acids during decomposition. Green manuring improves soil structure, increases water-holding capacity and decreases soil loss by erosion. Growing of green-manure crops in the off season reduces weed proliferation and weed growth. Green manuring helps in reclamation of alkaline soils. Root-knot nematodes can be controlled by green manuring.

5.6.7 Concentrated Organic Manures

Concentrated organic manures have higher nutrient content than bulky organic manure. The important concentrated organic manures are oilcakes, blood meal, fish manure etc. These are also known as organic nitrogen fertilizer. Before their organic nitrogen is used by the crops, it is converted through bacterial action into readily usable ammoniacal nitrogen and nitrate nitrogen. These organic fertilizers are, therefore, relatively slow acting, but they supply available nitrogen for a longer period.

5.6.8 Oil Cakes

After oil is extracted from oilseeds, the remaining solid portion is dried as cake which can be used as manure. The oil-cakes are of two types such as:

- **Edible oil cakes,** which can be safely fed to livestock *e.g.* Groundnut cake, Coconut cake etc., and
- **Non-edible oil cakes,** which are not fit for feeding livestock *e.g.* Castor cake, Neem cake, Mahua cake etc.,

Both edible and non-edible oil cakes can be used as manures. However, edible oil cakes are fed to cattle and nonedible oil cakes are used as manures especially for horticultural crops. Nutrients present in oil-cakes, after mineralization, are made available to crops 7 to 10 days after application. Oilcakes need to be well powdered before application for even distribution and quicker decomposition.

5.6.9 Other Concentrated Organic Manures

Blood-meal when dried and powdered can be used as manure. The meat of dead animals is dried and converted into meat-meal which is a good source of nitrogen (Table 5.6).

Table 5.6: Nutrient content of different concentrated manures

Organic Manure/s	Nutrient Content/s (%)		
	N	P_2O_5	K_2O
Blood meal	10-12	1-2	1.0
Meat meal	10.5	2.5	0.5
Fish meal	4-10	3-9	0.3-1.5
Horn and Hoof meal	13	-	-
Raw bone meal	3-4	20-25	-
Steamed bone meal	1-2	25-30	-

5.7 Organic Grazing

Grasslands have enormous potential for storing carbon (C) in the soil. Carbon sequestration improves soil health, makes soils more resilient to extreme weather events, contributes to climate change mitigation and can benefit pasture quality. In sustainable livestock grazing systems, the key challenge is to find the best type of management to combine animal production with soil ecosystem services such as carbon storage, nutrient cycling and biodiversity. Since 2005, dairy farmer Rob Richmond has been managing his pastures with a focus on building soil carbon. His grazing practices have helped him improve the organic matter levels in his soils. He uses the method of strip grazing, which confines the cattle to a limited area of grazing land for a short period, giving the animals a fresh allocation of pasture each day. Good pasture management leads to healthy soils.

5.7.1 Ways from Good Pasture Management: Besides building drought resistance, the following ways are used improvements in soil health occur from good pasture management:

- Ability of the soil to act as a filter, protecting water and air quality
- Better soil conditions for germination, seedling establishment, vegetative reproduction and root growth
- Carbon sequestration from air
- Improved water infiltration
- Increased plant production and reproduction

- Increased soil organic matter increases water available for plant growth
- More nutrients available for plant growth
- Reduced soil erosion from water

Some conservation practices for developing and maintaining healthy and productive pastures include (NOF, 2016):

- Establishment of desirable or more productive forages as needed. Some producers may include annual plantings to extend the grazing season, provide forages during seasons when perennials are typically less productive, or have animals deposit manure on cropland.
- Infrastructure and management practices may also be used to meet USDA organic requirements to protect environmentally sensitive areas in or adjacent to the pasture.
- Occasional mowing or mixed-species grazing to reduce weeds and stimulate growth of desirable species. Avoid prescribing forage species that require burning for germination or maintenance, or allow for alternatives such as grazing and tillage.
- Rotational grazing system and nutrient management appropriate for the climate and terrain. This is not particular to organic, but rotation is vital to pasture health and reduced environmental impacts, as required by NOP.
- Use of infrastructure such as animal trails and strategic locations of watering facilities, minerals, shade, and supplemental feed to support pasture rotation and utilization.

Overgrazing can lead to significant long-term degradation and an overall reduction in pasture condition and yields. Peterson says overgrazing is often the reason for diminished soil properties, which allows undesirable and invader species to gain a foothold.

5.7.2 Four basic Rules of Peterson: To avoid overgrazing, Peterson recommends graziers follow four basic rules (Johnson, 2019):

1. **Balance Stocking Rate with Forage Availability:** Determining initial stocking rate requires collecting information on overall pasture production and balancing the animal numbers with available forage.
2. **Increase Management Control by Increasing the Number of Paddocks:** Grazing period/rest period length are key factors in allowing plants adequate recovery time. Grazing periods that are too long and recovery periods that are too short are a major reason why pasture conditions deteriorate.

3. **Improve Utilization Rate:** Increasing the control over the livestock allows better utilization of all forage plants – not just the most palatable plants the livestock prefer. Peterson recommends the old standby of "take half, leave half," referring to how heavily an area is grazed. Grazing more than 50% stops all above- and below-ground plant growth.

4. **Lengthen Plant Rest and Recovery Time:** Depending on season of use, pastures should rest for at least 30-90 days after grazing. This allows for recovery of a diverse mixture of plants.

5.8 Livestock Organic Management

Livestock production has become far more sophisticated since the days of hunter-gatherers. Today, most production systems are intensive, with a very high per-animal productivity, due to better nutrition, health and housing management. However, the recent focus on and concerns over food quality, animal welfare, traceability, human health and environmental quality have led to the emergence of and growing interest in organic livestock farming, which is gradually spreading across the world. Tropical countries may find new opportunities in this premium market to export their own organic livestock products but it is a tough challenge, given the current sanitary conditions and prevalence of infectious diseases in these countries. Nevertheless, tropical countries have some natural advantages when it comes to organic livestock farming, which could be harnessed to boost organic livestock production for domestic consumption, initially, and exports in the long term. Nonetheless, this presents a formidable task, considering the stringent principles, guidelines, practices and standards of organic livestock production, as well as the mandatory certification procedures for such production systems. To benefit from this emerging system of food production, producers in developing countries must build their capacity and take into account their natural advantages. Presently, there are around 5,000 Farmer Producers' Organizations (FPOs) including FPCs in existence in the country, which were formed under various initiatives of the Central Government, State Governments, NABARD and other organizations over the last 8-10 years. FPOs are farmers' collective organizations, with membership mainly comprising small/marginal farmers (around 70 to 80%). For the success of FPOs in case of livestock products, the need is to create some model FPOs in the country, may be on the lines of the famous 'milk cooperative'(National Paper - PLP 2019-20). Livestock FPOs need support from the government to stand on their feet. Recently, many successful milk cooperatives in the country are surviving on their own while competing in the open market, but they have institutional support in the form of the National Dairy Development Board. Similarly, the government needs to create some other specific institutions for the benefit of FPOs created for production of livestock products other than milk. Formation of such SHGs and FPOs is more important in small ruminants farming, as there is complete absence of organizations of small ruminant's keepers. Formation of FPOs, cooperative societies or the SHGs based on the model demonstrated by the NDRI in dairy sector in villages, would enhance farmer's income and value addition. The producers should be linked to market. Further, clean milk production, traceability, adulterants detection kits and value addition of milk and milk products should be prioritized.

5.8.1 Status of Livestock

The livestock population in India has increased by 4.6%, from 512 million in 2012 to nearly 536 million in 2019. Though the overall number of cattle has increased only marginally, the cows, which account for over one-fourth of the total livestock population, showed an impressive increase of 18% over the previous census. Even among the indigenous (desi) breeds of cattle, female ones (cows) showed an increase of 10% in their number. Overall population of indigenous breeds of cattle (male and female), however, recorded a decline of 6% whereas population of cross-bred/exotic cattle recorded an increase of 27% over the 2012 census. The 20th livestock census figures shows that nearly 75% of total cattle in the country are female (cows) a clear sign of dairy farmers' preferences for milk-producing cattle. This also gained momentum in the past couple of years due to the government's assistance in terms of providing sex-sorted artificial insemination (AI), with semen of high-yielding bulls, free of cost at farmers› doorstep (Mohan, 2019).

- In 20th Livestock Census, 35.94%-Cattle, 27.80% Goat, 20.45% Buffaloes, 13.87% Sheep, 1.69% Pigs.
- Mithun, Yaks, Horses, Ponies, Mules, Donkeys, and Camels taken together contribute 0.2% of the total livestock.
- As compare to previous census the percentage share of sheep and goat population has increased whereas the percentage share of cattle, buffalo and pigs has marginally declined.

Table 5.7: Percentage sharing of Livestock

Category	Population in Million (2012)	Population in Million (2019)	% Growth
Cattle	190.90	192.9	0.83
Buffalo	108.70	109.85	1.06
Sheep	65.07	74.26	14.1
Goat	135.17	148.88	10.1
Pig	10.29	9.06	-12.03
Mithun	0.30	0.38	26.66
Yak	0.08	0.06	-25.00
Horses & Ponies	0.63	0.34	-45.58
Mule	0.20	0.08	-57.09
Donkey	0.32	0.12	-61.23
Camel	0.40	0.25	-37.05

Source: Mohan, (2019)

5.8.2 Role of Livestock in Organic Farming

5.8.2.1 Nutrient Cycling: Nitrogen fixed by leguminous plants and different nutrients devoured by farm animals amid brushing are come back to soil through dung and urine. Overseen pain stakingly, farm animals and manures can assume an imperative part in nutrient cycling on the organic farming. In feedlots, it is important to store and discard manure and urine in a naturally acceptable way. Excreta contain several nutrients (including nitrogen, phosphorus and potassium) and organic matter, which are important for maintaining soil structure and fertility. Stubble in the fields and crop residues are important sources of forage in smallholder systems. Lower mature leaves stripped from standing crops, plants thinned from cereal stands and vegetation on fallow fields offer additional fodder resources related to food cropping. When animals consume vegetation and produce dung, nutrients are recycled more quickly than when the vegetation decays naturally. Grazing livestock transfer nutrients from range to cropland and concentrate them on selected areas of the farm.

5.8.2.2 Pasture and Crop Establishment: Farm animals can help with planning the ground for planting. For instance, they can help with stubble management by brushing and trampling the stubble.

5.8.2.3 Providing Energy: Excreta is the basis for the production of biogas and energy for household use (e.g. cooking, lighting) or for rural industries. Fuel in the form of biogas or dung cakes can replace charcoal and wood.

5.8.2.4 Insect and Disease Control: Fodder part of the mixed cropping system builds a critical fertility and structure building phase into rotations and intrudes the potential for the development of insects and disease.

5.8.2.5 Land Preparation for Cropping: Farm animals such as pigs can 'plough' rough or new land before planting vegetables or grains, reducing tillage and weed control costs.

5.8.2.6 Soil Water Storage: It results in greater soil water storage capacity, mainly because of biological aeration and the increase in the level of organic matter.

5.8.2.7 Weed Control: Farm animals are utilized widely for weed control on natural farming. For example, they can graze down weeds either before sowing a crop or after crop establishment for weed control and to enhance tillering. Crops can be selected for their palatability. Farm animals specifically touch out weeds and keep away from the less palatable fodder.

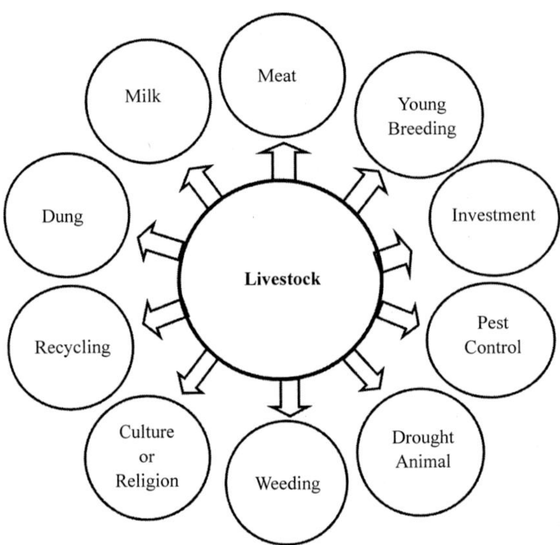

Fig 5.6: Livestock Organic Management

If supported by successive capacity and knowledge building taken as a strategy and the establishment of certifying organizations and promotion of the importance of organic farming to initiate consumers awareness of organic products, its nature of being environmental friendly and capacity to maintain the soil fertility and integrity, its help of farmers on control over their means of production and greater independence will help more farmers of developing countries to initiate and engage in this important farming practice and contribute to the wellbeing of the environment, the livestock species, the human being in general. Therefore, organic farming is the way towards sustainable development for developing countries (Daniel, 2016; Namdhini and Suganthi, 2018).

5.9 Zero Budget Natural Farming in India

Zero Budget Natural Farming (ZBNF), which is a set of farming methods, and also a grassroots peasant movement, has spread to various states in India. It has attained wide success in southern India, especially the southern Indian state of Karnataka where it was first evolved. A rough estimation for just Karnataka puts the figure there at around 100,000 farmer families, while at the national level, ZBNF leaders claim that numbers could run into millions. This has been achieved without any formal movement organization, paid staff or even a bank account. Indian farmers increasingly find themselves in a vicious cycle of debt, because of the high production costs, high interest rates for credit, the volatile market prices of crops, the rising costs of fossil fuel based inputs, and private seeds. More than a quarter of a million farmers have committed suicide in India in the last two decades. Various studies have linked farmer's suicides to debt. Debt is a problem for farmers of all sizes in India. Under such conditions, 'zero budget' farming promises to end a reliance on loans and drastically cut production costs, ending the debt cycle for desperate farmers. The word 'budget' refers to credit and expenses, thus the phrase 'Zero Budget' means without using any credit, and without spending any money on purchased inputs. 'Natural farming' means farming with Nature and without chemicals. The father of ZBNF and Padma Shri Awardee, Sh. Subash Palekar has provided four pillars of ZNBF (Fig 5.7).

5.9.1 Pillars of ZNBF

- Jiwamrita nectar of life (consisting of microbes) that is prepared from dung and urine of indigenous Kapila cow (not other animals like exotic or cross-bred cows, bulls or buffaloes).
- Bijamrita (Seed Treatment using local cowdung and cow urine).
- Acchadana - Mulching (activities to ensure favorable microclimate in the soil), and
- Waaphasa (soil aeration).

5.9.1.1 Jivamrita/Jeevamrutha is a fermented microbial culture. It provides nutrients, but most importantly, acts as a catalytic agent that promotes the activity of microorganisms in the soil, as well as increases earthworm activity; During the 48hrs fermentation process, the aerobic and anaerobic bacteria present in the cow dung and urine multiply as they eat up organic ingredients (like pulse flour). A handful of undisturbed soil is also added to the preparation, as inoculate of native species of microbes and organisms. Jeevamrutha also helps to prevent fungal and bacterial plant diseases.

5.9.1.2 Bijamrita/Beejamrutha is a treatment used for seeds, seedlings or any planting material. Bijamrita is effective in protecting young roots from fungus as well as from soil-borne and seed-borne diseases that commonly affect plants after the monsoon period. It is composed of similar ingredients as jeevamrutha - local cow dung, a powerful natural fungicide, and cow urine, a strong anti-bacterial liquid, lime, soil.

5.9.1.3 Acchadana - Mulching. There are three types of mulching such as

- **Soil Mulch:** This protects topsoil during cultivation and does not destroy it by tilling. It promotes aeration and water retention in the soil. In this practice deep ploughing may be avoided.
- **Straw Mulch:** Straw material usually refers to the dried biomass waste of previous crops, but it can be composed of the dead material of any living being (plants, animals, etc).
- **Live Mulch** (symbiotic intercrops and mixed crops): It is essential to develop multiple cropping patterns of monocotyledons (monocots; Monocotyledons seedlings have one seed leaf) and dicotyledons (dicots; Dicotyledons seedlings have two seed leaves) grown in the same field, to supply all essential elements to the soil and crops. For instance, legumes are of the dicot group and are nitrogen-fixing plants. Monocots such as rice and wheat supply other elements like potash, phosphate and sulphur.

5.9.1.4 Whapasa-moisture: Whapasa is the condition where there are both air molecules and water molecules present in the soil, and he encourages reducing irrigation, irrigating only at noon, in alternate furrows ZBNF farmers report a significant decline in need for irrigation in ZBNF.

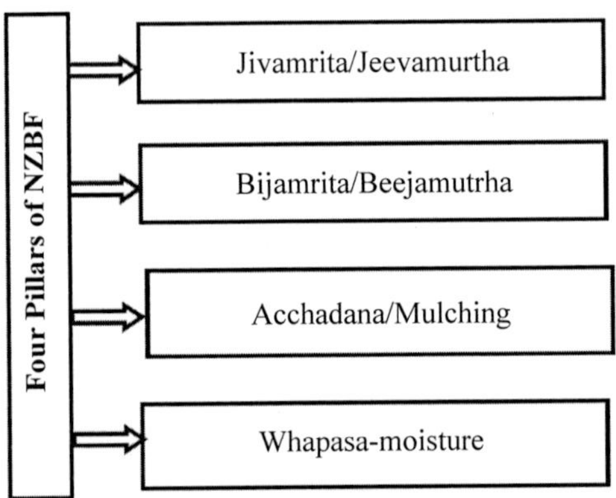

Fig. 5.7: Four pillars of ZNBF

5.9.2 Other important principles of ZBNF and points to note

a) Intercropping.

b) Contours and bunds

c) Local species of earthworms.

d) Cow dung

In addition, ZBNF includes three methods of insect and pest management: Agniastra, Brahmastra and Neemastra (all different preparations using cow urine, cow dung, tobacco, fruits, green chilli, garlic and neem).

ZBNF has recently been re-christened as 'Subhash Palekar Natural Farming' (SPNF) after the name of its prime mover, Mr Subhash Palekar, who taught this new method of farming to the farmers in his 1st workshop that was organized at Wayanad in Kerala in the year 2008 (http://palekarzerobudgetspiritualfarming.org). But the system is supposed to have existed since mid 1990s in southern India, notably in the state of Karnataka. Later, it was promoted by the Government of Andhra Pradesh during the last two decades as 'Community Managed Sustainable Agriculture' (CMSA) or 'Climate Resilient Zero Budget Natural Farming' (CRZBNF). The Government of Andhra Pradesh spent about` 324cr. to popularize it among the farmers and proposed to extend it to six million farmers of the state cultivating 9Mha area by 2022-24 in a phased manner. The move to turn Andhra Pradesh into world's 1st natural farming state was estimated to cost` 17,000cr. (equivalent to 2.3 billion US $) and was proposed to be raised as loan based on state guarantee. 'Azim Premji Philanthropic Initiative' has already extended a` 100 crore grant for this purpose. Apart from this, attempts are being made to promote this new method of farming in a big way by non-government organizations as well as governmental agencies as presented below:

5.9.3 Endorsement by the Government of India

While the country has been planning to revamp its agricultural production system including R&D to meet the formidable challenges being faced by it, the economic survey of 2018-19 made a fervent appeal for the adoption of 'Zero Budget Natural Farming' (ZBNF) in a big way to double farmers' income and it was subsequently endorsed by the Hon'ble Finance Minister during her budget speech in the parliament. ZBNF is said to be 'Zero Cost or Zero Input Natural Farming' and, therefore, whatever quantity is harvested is treated as net profit to the farmer.

5.9.4 Need for Scientific Validation

There have been a few reviews and opinions on ZBNF (Munster, 2018; Agarwal, 2019; Ramakumar and Arjun, 2019 and an editorial in 'Economic and Political Weekly' 'Mirage of Zero Budget Farming' in July 27, 2019 issue). One of the leading journalists, Mr Vivian Fernandes recently visited Palekar's farm and interacted with the representative of Mr Palekar at his farm as well as farmers near Palekar's farm. The article "Does Zero Budget Natural Farming Work? A Reality Check" appeared in 15th July 2019 edition of the 'Quint'. All the above mentioned articles conclude that the claims of ZBNF are questionable and need further scientific validation. In fact, a considerable number of farmers in Maharashtra have shifted to chemical nutrient-based farming (Bhosale, 2019).

5.10 Tillage

"Tillage" can also mean the land that is tilled. The word "cultivation" has several senses that overlap substantially with those of "tillage". In a general context, both can refer to agriculture. Within agriculture, both can refer to any kind of soil agitation. Additionally, "cultivation" or "cultivating" may refer to an even narrower sense of shallow, selective secondary tillage of row crop fields that kills weeds while sparing the crop plants.

5.10.1 Kinds of Tillage

5.10.1.1 Conservation Tillage: Any tillage and planting system in which at least 30% of the soil surface is covered by plant residue after planting to reduce soil erosion by water; or where soil erosion by wind is the primary concern of at least 1,000 pounds per acre of flat small grain residue equivalent are on the surface during the critical erosion period.

5.10.1.2 Conventional Tillage: Tillage types that leave less than 15% residue cover after planting, or less than 500 pounds per acre of small grain residue equivalent throughout the critical wind erosion period. Generally involves plowing or intensive tillage.

5.10.1.3 Reduced Tillage: Tillage types that leave 15-30% residue cover after planting or 500 to 1,000 pounds per acre of small grain residue equivalent throughout the critical wind erosion period.

5.10.2 Modern Concepts of Tillage

In conventional tillage, energy is often wasted and sometimes, soil structure is destroyed. Recently considerable changes has taken place in tillage practices and several new concepts have been introduced namely, minimum tillage, zero tillage, stubble mulch tillage. The immediate cause for introducing minimum tillage was high cost of tillage due to steep rise in oil prices. In addition there are problems associated with conventional tillage. Repeated use of heavy machinery, destroys structure, causes soil pans and leads to erosion. The needs of planting zone (row zone) and water management zone (inter row zone) are different. In row crops, it is sufficient to provide fine tilth in the row zone for creating conditions optimal for sowing and conducive to rapid and complete germination and seedling establishment. In the inter-row zone, secondary tillage is not done and it should be rough and cloddy where soil structure is coarse and open so that weeds may not germinate and more water infiltrates into the soil. The important object of tillage is weed control which can be done by herbicides. The Practice of inverting the top soil in order to bury manures and crop residues becomes less important object of tillage in modem field management as the use of animal and green manure is rather uncommon. Crop residues can and in many cases should be left over the surface as stubble mulch to protect against evaporation and erosion losses. Research has shown that frequent tillage is rarely beneficial and often detrimental. All these reasons led to the development and practice of (a) minimum tillage, (b) zero tillage and (c) stubble mulch farming etc.

5.10.2.1. Minimum Tillage

It involves considerable soil disturbance, though to a much lesser extent than that associated with conventional tillage. Minimum tillage is aimed at reducing tillage to the minimum necessary for ensuring a good seedbed, rapid germination, a satisfactory stand and favourable growing conditions. Minimum tillage can be reduced in two ways such as (a) by omitting operation which do not give much benefit when compared to the cost, and (b) by combining agricultural operations like seeding and fertilizer application. Some different methods of minimum tillage are practiced such as

- **Row Zone Tillage**: After primary tillage with mould board plough, secondary tillage operations like disking and harrowing are reduced. The secondary tillage is done in the row zone only.
- **Plough Plant Tillage**: After the soil is ploughed, a special planter is used and in one run over the field, the row zone is' pulverized and seeds are sown.
- **Wheel Track Planting**: Ploughing is done as usual. Tractor is used for sowing and the wheels of the tractor pulverize the row zone.

Advantages of Minimum Tillage

- Improved soil conditions due to decomposition of plant residues in situ;
- Higher infiltration caused by the vegetation present on the soil and channels formed by the decomposition of dead roots;
- Less resistance to root growth due to improved structure;
- Less soil compaction by the reduced movement of heavy tillage vehicles and less soil erosion compared to conventional tillage.

Disadvantages of Minimum Tillage

- Continuous use of herbicides cause pollution problems and dominance of perennial problematic weeds.
- In minimum tillage, more nitrogen has to be added as rate of decomposition of organic matter is slow.
- Nodulation is affected in some leguminous crops like peas and broad beans.
- Seed germination is lower with minimum tillage.
- Sowing operations are difficult with ordinary equipment.

5.10.2.2 Stubble Mulch Tillage: Conventional method of tillage results in soil erosion. Stubble mulch tillage or stubble mulch farming a new approach was developed for keeping soil protected at all times whether by growing a crop or by crop residues left on the surface during fallow periods. It is a year round system of managing plant residue with implements that undercut residue, loosen the soil and kill weeds. Sweeps or blades are generally used to cut the soil up to 12 to 15cm depth in the first operation after harvest and the depth of cut reduced during subsequent operations. When unusually large amount of residues are present, a disc type implement is used for the first operation to incorporate some of the residues into the soil. This hastens decomposition, but still keeps enough residues on the soil.

5.10.2.3 Zero Tillage: Zero tillage is the process where the crop seed will be sown through drillers without prior land preparation and disturbing the soil where previous crop stubbles are present. Zero tillage not only reduce the cost of cultivation it also reduces the soil erosion, crop duration and irrigation requirement and weed effect which is better than tillage. Zero Tillage (ZT) is also called No Tillage or Nil Tillage.

5.10.3 Zero Tillage in India

The zero-tillage system is being followed in the Indo-Gangetic plains where rice-wheat cropping is present. Wheat will be planted after rice harvest without any operation. Hundreds of farmers are following the same system and getting more yields and profits by reducing the cost of cultivation. In South, the southern districts like Guntur and some parts of West Godavari of Andhra Pradesh state follow the ZT system in rice-maize cropping system. The green revolution paved the way for the rice-wheat production system in the north-western parts of India. But in due course of time, the yields of rice and wheat become stagnant due to inappropriate soil and water management system and late planting of wheat, as in the hot season rice is being grown and in the winter wheat follows the rice. In 1990's the zero tillage came to mitigate the problem, by planting the wheat by drilling without any land preparation and tillage. ZT proves better for direct-seeded rice, maize, soybean, cotton, pigeonpea, mungbean, clusterbean, pearlmillet during kharif season and wheat, barley, chickpea, mustard and lentil during rabi season. Wheat sowing after rice can be advanced by 10-12 days by adopting this technique compared to conventionally tilled wheat, and wheat yield reduction caused by late sowing can be avoided. ZT provides opportunity to escape wheat crop from terminal heat stress. ZT reduces cost of cultivation by nearly Rs 2,500-3,000/ha through reduction in cost of land preparation, and reduces diesel consumption by 50-60 litres/ha. ZT reduces water requirement of crop and the loss of organic carbon by oxidation. Zero tillage reduces *Phalaris minor* problem in wheat. The carbon status of soil is significantly enhanced in surface soil (0-5cm), particularly under crop residue retention with zero tillage (Sendhil *et al.,* 2018).

5.10.3.1 Advantages of Zero Tillage

- Reduction in the crop duration and thereby early cropping can be obtained to get higher yields.
- Reduction in the cost of inputs for land preparation and therefore a saving of around 80%.
- Residual moisture can be effectively utilized and number of irrigations can be reduced.
- Dry matter and organic matter get added to the soil.
- Environmentally safe - Greenhouse effect will get reduced due to carbon sequestration.
- No tillage reduces the compaction of the soil and reduces the water loss by runoff and prevent soil erosion.
- As the soil is intact and no disturbance is done, No Till lands have more useful flora and fauna.

References

Ali, S.; Nawaz, A.; Syed Asad Hussain Bukhari; Ejaz, S.; Ahmad, S. (2019). Integrated Pest and Disease Management for Better Agronomic Crop Production Agronomic Crops pp. 385-428.

Ameena, M.; Kumari, V.L.G.; George, S. (2006). Indian J. Weed Sci. 38: 81-85.

Asante, S. K.; Tamo, M.; Jackai, L.E.N. (2001). Afr. Crop Sci. J. 9: 655–665.

Bhosale, J. (2019) Zero Budget farming has few takers in the state where it originated retrieved from https://economictimes.indiatimes.com/news/economy/agriculture/budget-2019-zero-budget-farming-has-few-takers-in-the-state-where-it-originated/articleshow/70089472.cms?from=mdr

Brummer, E.C. (1998). Agronomy Journal 90:1–2.

Chandrakar, B.L.; Urkurkar, J.S. (1993). Indian J. Weed Sci. 25: 32-35.

Commission Regulation (EC) No 889/2008 of 5 September 2008 laying down detailed rules for the implementation of Council Regulation (EC) No 834/2007 on organic production and labelling of organic products with regard to organic production, labelling and control Official Journal of European Union L 250; 2008. 51. 1-84.

Curl, E. A. (1963). The Botanical Review 29: 413–479.

Daniel T; addesse Wolde; Tamir, B. (2016). Global Veterinaria 16 (4): 399-412.

Dara, S. K. (2018a). CAPCA Adviser 21: 44–50.

Dara, S.K. (2017). Insect resistance to biopesticides. University of California Agriculture and Natural Resources eJournal Strawberries and Vegetables. https://ucanr.edu/blogs/blogcore/postdetail.cfm?postnum=25819

Dara, S.K. (2019). Journal of Integrated Pest Management 10(1): 1–9.

Dara, S.K.; Peck, D.; Murray, D. (2018). Insects 9: 156.

Davis, J.L.; Armengaud, P.; Larson, T.R.; Graham, I.A.; White, P.J.; Newton, A.C.; Amtmann,A. (2018). Cell Environ. 41: 2357–2372.

Dodia, D.A.; Patel, I.S.; Patel, G.M. (2010). Botanical pesticides for pest management. Scientific Publishers (India), Jodhpur, India.

Dong, L.Q.; Zhang, K.Q. (2006). Plant Soil 288: 31–45.

Dosdall, L.M.; Herbut, M. J.; Cowle, N.T.; Micklich. T.M. (1996). Can. J. Plant Sci. 76: 169–177.

Douglas, A.E. (2018). Ann. Rev. Plant Biol. 69: 637–660.

El-Sayed, A.M., Suckling, D. M.; Byers, J.A.; Jang, E.B.; Wearing, C.H. (2009). J. Econ. Entomol. 102: 815–835.

Entz, M.H.; Baron, V.S.; Carr, P.M.; Meyer, D.W.; Smith, S.R.; Jr, and McCaughey, W.P. (2002). Agronomy Journal 94:240–250.

Entz, M.H.; Bullied, W.J.; Katepa-Mupondwa, F. (1995). Journal of Production Agriculture 8: 521–529.

Foster, S.P.; Harris, M.O. (1997). Annu. Rev. Entomol. 42: 123–146.

Gamliel, A.; Katan. J. (2012). American Phytopathological Society, St. Paul, MN.

Gogo, E.O.; Saidi, M.; Ochieng, J.M.; Martin, T; Baird, V.; Ngouajio, M. (2014). Hort. Science 49: 1298–1304.

Gogoi, A.K.; Rajkhowa, D.J.; Kandali, R. (2001). Indian J. Weed Sci. 33: 18-21.

Hajek, A.E.; Eilenberg, J. (2018). Natural enemies: An Introduction to Biological Control. Cambridge University Press, Cambridge, United Kingdom.

Heimpel, G.E.; Cock, M.J.W. (2018). Bio Control 63: 27–37.

Heinz, K.M., Parrella, M.P.; Newman, J.P. (1992). J. Econ. Entomol. 85: 2263–2269.

Hodson, A.K.; Lampinen, B.D. (2018). Arthropod Plant Interact. 1–11.

http://agricoop.nic.in/guidelines/integrated-nutrient-management

http://agritech.tnau.ac.in/org_farm/orgfarm_manure.html
http://subhamulticonsultancy.blogspot.com/
http://www.fao.org/agriculture/crops/core-themes/theme/spi/iclsd/en/
https://timesofindia.indiatimes.com/india/livestock-population-up-by-4-6-cow-count-rises-by-18/articleshow/71622775.cms
https://www.ctc-n.org/technologies/integrated-nutrient-management
https://www.nabard.org/auth/writereaddata/CareerNotices/2708183505Paper%20on%20FPOs%20-%20Status%20&%20%20Issues.pdf
https://www.nrcs.usda.gov/wps/portal/nrcs/ia/newsroom/features/25750f1b-1d0c-4abc-bc73-db796e3f44e1/
https://www.sciencedirect.com/topics/agricultural-and-biological-sciences/integrated-weed-management
https://www.thehindu.com/sci-tech/Enriched-farmyard-manure/article15516804.ece
Huang, F.; Andow, D.A.; Buschman, L.L. (2011). Entomol. Exp. Appl. 140: 1–16.
IRAC (Insecticide Resistance Action Committee). (2018). IRAC mode of action classification scheme. https://www.irac-online.org/documents/moa-classification/?ext=pdf
Itnal, C.J.; Lingaraju, B.S.; Kurdikeri, C.B. (1993). Indian J. Weed Sci. 25: 27-31.
Jayasooriya, H.J.C.; Aheeyar, M.M.M. (2016). Procedia Food Sci. 6: 208–212.
John, R.; Hendrickson, J.D.; Hanson, Donald L. Tanaka; Sassenrath, G. (2008). Renewable Agriculture and Food Systems: 23(4): 265–271
Johnson, J. (2019). Soil Health Important for Grazing Operations, Too retrieved from https://www.nrcs.usda.gov/wps/portal/nrcs/ia/newsroom/features/25750f1b-1d0c-4abc-bc73-db796e3f44e1/
Kakraliya, S.K.; Jat, R.D.; Kumar, S.; Choudhary, K.K; Prakash, J.; Singh, L.K. (2017). Int. J. Curr. Microbiol. Applied Sci. 6 (3): 152-163.
Karungi, J., Adipala, E.; Ogenga-Latigo, M.W.; Kyamanywa, S.; Oyobo, N. (2000). Crop Prot. 19: 231–236.
Kenis, M., Hurley, B.P.; Hajek, A.E.; Cock, M.J.W. (2017). Biol. Invasions 19: 3401–3417.
Klassen, W., Curtis. C.F. (2005). History of the sterile insect technique, pp. 3–36. In V. A. Dyck, J. Hendrichs, and A. Robinson (eds.), Sterile insect technique. Springer, Dordrecht, The Netherlands.
Kolhe, S.S. (2001). Indian J. Weed Sci.33: 26-29.
Krall, J.M.; Schuman, G.E. (1996). J. Production Agril. 9:187–191.
Kulmi, G.S.; Tiwari, P.N. (2004). Indian J. Weed Sci. 36: 104-107.
Kulmi, G.S.; Tiwari, P.N. (2005). Indian J. Weed Sci. 37: 77-80.
Kulmi, G.S.; Tiwari. P.N. (2005). Indian J. Weed Sci. 37: 77-80.
Kumar, S. Angiras, N.N.; Singh, R. (2006). Indian J. Weed Sci. 38: 73-76.
Kumar, V.; Ramjan, Md.; Das, T. (2019). Biomolecule Reports- An International eNewsletter BR/01/18/05
Kunjwal, N.; Srivastava, R.M. (2018). Insect pests of vegetables, pp. 163– 221. In Omkar (ed.), Pests and their management. Springer, Singapore.
Lacey, L.A. (2017). Microbial control of insect and mite pests: from theory to practice. Academic Press, London, United Kingdom.
Larkin, R.P. (2008). Soil Biol. Biochem. 40: 1341–1351.
Lasota, J.A.; Dybas, R.A. (1991). Annu. Rev. Entomol. 36: 91–117.
Leach, H.J.; Moses, E.; Hanson, P.; Fanning, Isaacs, R. (2017). J. Pest Sci. 91: 219–226.
Lefebvre, M., Langrell, S.R.H.; Gomez-y-Paloma, S. (2015). Incentives and policies for integrated pest management in Europe: a review. Agron. Sustainable Dev. 35: 27–45.
Liebman, M., Dyck. E. (1993). Ecol. Appl. 3: 92–122.
López-Bucio, J. Pelagio-Flores, R.; Herrera-Estrella, A. (2015). Sci. Horticult. 196: 109–123.

Mankau, R. (1981). Microbial control of nematodes, pp. 475–494. In B. Zuckerman and R. A. Rohde (eds.), Plant parasitic nematodes, Vol. III. Academic Press, New York, NY.
Mathew, G.; Sreenivasan, E.; Mathew, J. (1995). Indian J. Weed Sci. 27: 42-44.
Merfield, (2019). Integrated Weed Management in Organic Farming. Organic Farming Global Perspectives and Methods pp. 117-180.
Mitchell, C.E.; Reich, P.B.; Tilman, D.; Groth, J.V. (2003). Global Change Biol. 3: 438–451.
Mohan, V. (2019) Livestock population up by 4.6%; cow count rises by 18%. Retrieved from https://timesofindia.indiatimes.com/india/livestock-population-up-by-4-6-cow-count-rises-by-18/articleshow/71622775.cms
Mohler, C. L.; Johnson, S.E. (2009). Crop rotation on organic farms a planning manual. Natural Resource, Agriculture, and Engineering Service, Ithaca, NY.
Mondal, R.; Goswami, S.; Mondal, K.; Goswami, S.B.; Jana, K. (2019). Progressive Agricultural Sciences 1(1): 89-97.
Morrison, W.R.; Lee, D.-H.; Short, B.D.; Khrimian, A.; Leskey, T.C. (2016). J. Pest Sci. 89: 81–96.
Nagar, R.K.; Meena, B.S.; Dadheech, R.C. (2009). Indian J. Weed Sci. 41: 71-75.
Nandal, T.R.; Singhal, R. (2002). Indian J. Weed Sci. 34: 72-75.
Nandhini, U.D.; Suganthi, S. (2018). Livestock Importance in Organic Farming Approaches in Poultry, Dairy & Veterinary Sciences Vol 1 Issue 1 DOI: 10.31031/APDV.2018.05.000602
National Organic Farming Handbook 2016 United States Department of Agriculture (USDA). 190–612–H, 2nd Ed., Feb 2016 retrieved from https://directives.sc.egov.usda.gov/OpenNonWebContent.aspx?content=39107.wba
National Paper - PLP 2019-20 Farmer Producers' Organizations (FPOs): Status, Issues and Suggested Policy Reforms https://www.nabard.org/auth/writereaddata/Career Notices/2708183505Paper%20on%20FPOs%20-%20Status%20&%20%20Issues.pdf
Nielsen, A. L.; Dively, G.; Pote, J.M.; Zinati, G.; Mathews, C. (2016). Environ. Entomol. 45: 472–478.
Parsa, S.; Morse, S.; Bonifacio, A.; Chancellor, T.C.B.; Condori, B.; Crespo-Pérez, V.; Hobbs, S.L.A.; Kroschel, J.; Ba, M.N.; Rebaudo, F. et al. . (2014). Proc. Natl. Acad. Sci. U.S.A. 11: 3889–3894.
Paulitz, T.C.; Bélanger, R.R. (2001). Annu. Rev. Phytopathol. 39: 103–133.
Peterson, R.K.D.; Higley, L.G.; Pedigo, L.P. (2018). Am. Entomol. 64: 146–150.
Pimentel, D.; McNair, S.; Janecka, J.; Wightman, J.; Simmonds, C.; O"Connell, C. (2001). Agriculture, Ecosystems & Environment 84(1): 1-20.
Porwal, M.K. (1995). Indian J. Weed Sci. 27: 16-18.
Poston, F.L.; Pedigo, L.P.; Welch, S.M. (1983). Am. Entomol. 29: 49–53.
Pretty, J.; Bharucha, Z.P. (2015). Insects. 6: 152–182.
Rameshwar, S.C.; Sharma, G.D.; Rana, S. (2002). Indian J. Weed Sci. 34; 68-71.
Rana, S.S. (2016). Organic Farming Department of Agronomy, COA, CSK HPKV, Palampur, HP.
Rao, A.N.; Nagamani, A. (2007). Available technologies and future research challenges for managing weeds in dry-seeded rice in India. In: Proc. of the 2151 Asian-Pacific Weed Science Society Conf. 2nd to 6th October 2007, Colombo, Sri Lanka.
Rao, A.N.; Nagamani, A. (2010). Indian J. Weed. Sci. 42: 1-10.
Rathi, J.P.S.; Tewari, A.N.; Kumar, M. (2004). Lndian -X-Weed Sci. 36: 218-220.
Rezaei, R.; Safa, L.; Damalas, C.A.; Ganjkhanloo, M.M. (2019). J. Environ. Manage. 236:328–339.
Roy, R. N.; Finck, A.; Blair, G.J.; Tandon, H.L.S (2006) Plant nutrition for food security, FAO Rome.
Sardana, V., Singh,S.; Sheoran, P. (2006). Indian J. Weed Sci. 38: 77-80.

Sarfraz, M.; Dosdall, L.M.; Keddie, B.A. (2005). Outlooks on Pest Management 16: 78–84.
Schiere, J.B.; Ibrahim, M.N.M.; van Keulen, H. (2002). Agricultural Ecosystems and Environment 90:139–153.
Sharma, H.S.; Fleming, C.; Selby, C.; Rao, J.R.; Martin, T. (2014). J. Appl. Phycol. 26: 465–490.
Shekara, B.G.; Nanjappa, H.V. (1993). Indian J. Weed Sci. 25: 40-43.
Shinde, S.H., Pawar, V.S.; Suryawanshi, G.B.; Ahire, N.R.; Surve, U.S. (2003). Indian J. Weed Sci. 35:90-92.
Shorey, H.H.; Gerber, R.G. (1996). Environ. Entomol. 25: 1154–1157.
Singh, C.M.; Sharma, P.K; Kishor, P.; Mishra, P.K.,; Singh, A.P.; Verma, R.; Raha, P. (2011). Asian J. Agril. Res 5(1): 76-82.
Singh, P.K.; Prakash, O.; Singh, B.P. (2001). Indian J. Weed Sci. 33: 139-142.
Singh, P.K.; Prakash, O.; Singh, B.P. (2001c). Ind. J. Weed Sci. 33: 139-142.
Singh, R. (2006). Indian J. Weed Sci. 38: 69-72.
Singh, R.; Singh, B. (2004). Indian J. Weed Sci. 36: 25-27.
Singh, R.; Singh, B.; Chauhan, K.N.K. (1999). Indian J. Weed Sci. 31:138-141.
Singh, S.N.; Singh, R.K.; Singh, B. (2001a). Indian J. Weed Sci. 33: 136-138.
Singh, S.P.; Kumar, R.M. (1999). Indian J. Weed Sci.31: 222-224.
Singh, V.P.; Singh, G. (2001). Indian J. Weed Sci. 33: 52-55.
Snoeijers, S. S.; Pérez-García, A.; J. Joosten, M.H.A.; De Wit., P.J.G.M. (2000). Eur. J. Plant Pathol. 106: 493–506.
Sparks, T.C.; Nauen, R. (2015). Pestic. Biochem. Physiol. 121: 122–128.
Stenberg, J. A. (2017). Trends Plant Sci. 22: 749–769.
Sukhadia, N.M.; Ramani, B.B.; Dudhatra, M.G. (2002). Indian J. Weed Sci. 34: 76-79.
Tabashnik, B. E.; Brévault, T.; Carrière, Y. (2013). Nat. Biotechnol. 31: 510–521.
Tang, J.D., Collins, H.L.; Metz, T.D.; Earle, E. D.; Zhao, J.Z.; Roush, R.T.; Shelton, A.M. (2001). J. Econ. Entomol. 94: 240–247.
Termorshuizen, A.J. (2002). In "Plant Pathologist's Pocketbook", pp. 318-327, eds, J.M. Waller, J.M. Lenné and S.J. Waller. (CAB International, Wallingford, UK).
Tewari, A.N.; Tiwari, S.N.; Rathi., J.P.S.; Singh, B.; Tripathi, A.K. (2003). Indian J. Weed Sci. 35: 49-52.
TNAU Portal , http://agritech.tnau.ac.in/
van Lenteren, J.C. (1988). Annu. Rev. Entomol. 33: 239–269.
Vijaykumar, B.; Reddy, M.N.; Shivashankar, M. (1995). Indian J. Weed Sci. 27: 12-15.
Vladés, E.M.A.E.; Aldana, L.L.L; Figueroa, B.R.; Gutiérrez, O.M.; Hernández, R.M.C.; Chavelas, M.T. (2005). Fla. Entomol. 88: 338–340.
Webb, S.E.; Linda, S.B. (1992). J. Econ. Entomol. 85: 2344–2352.
Wijnands, F.G.; Sukkel, W.; Booij, C. (2000). In "Biologisch bedrijf onder de loep: Biologische akkerbouw en vollegrondsgroenteteelt in perspectief", pp. 65-71, eds. F.G. Wijnands, J.J. Schröder, W. Sukkel and R. Booij. (Praktijk Onderzoek Plant en Omgeving, Lelystad, The Netherlands).
Wright, R.J. (1984). J. Econ. Entomol. 77: 1254–1259.
Zalom, F.G.; Bolda, M.P.; Dara, S.K.; Joseph. S. (2018). UC IPM pest management guidelines: strawberry (insects and mites). University of California Statewide IPM Program, Oakland, CA, Publication Number 3468.

6

Indian Farming and Doubling of Farmer's Income

6.1 Introduction

India experienced an impressive growth and productivity gains in agriculture since Independence reflecting our farmers' resilience against multiple odds and challenges. Despite a structural transformation post-Green Revolution, characterized by a consumer centric and food security policy objective, the regular distress and crises in the recent past pose a severe threat to income and livelihood security of farmers. The income realised from agriculture compared to nonagriculture, has been too low with a wide variation across regions. The average income of an agricultural household during 2012-13 has been estimated as low as at `6426 per month, marginally higher than the average consumption expenditure (`6223). Further, around 23 per cent of the farmers still live below the poverty line (DFI Committee, 2017). Estimates from the Central Statistics Office's (CSO) indicated that the annual growth in gross value added from agriculture has reduced drastically from as high as 6.9% (October-December, 2016) to a mere 2.3% (April-June, 2017) at 2011-12 prices in the recent past ten quarters. Shockingly, the growth was lower in nominal terms in comparison to real prices during the April-June, 2017 quarter indicating the creeping deflation in farm sector. In addition, the real farm revenue for a basket of pulses and vegetables estimated by the price multiplied with the quantity of market arrivals and deflated by the rural consumer price index has declined indicating the distress (Damodaran, 2017) and the difficulty it poses for the doubling objective within the stipulated period. Farmers are at the epicentre of Indian economy and their livelihood upliftment is a step towards holistic development of the nation. Decline in productivity and income has a serious implication on rural household poverty, and other economic, social as well as sustainability indicators (Timmer, 1995; Datt and Ravallion, 1998; Fan *et al.,* 1999; Mellor, 1999; Irz *et al.,* 2001; Minten and Barrett, 2008; Byerlee *et al.,* 2009; Muyanga, *et al.,* 2010). Hence, increasing the income of farmers from different sources across holding size and region has become an utmost priority for the policy planners. Though, the state goes rhetoric about farmers' welfare since independence, its policies have always been consumer centric preventing the producers from relishing the fruits of their labour and hard work. The agriculture policies have led to a 'boom-bust' cycle in agriculture with a certain regularity that a year of drought leads to prices shoot up, area increase, abundant production, collapse of prices, area shrinkage and prices shoot up.

6.2 Status of Household Income of Farmers

The realization of doubling the income of farmers comprise both farm and non-farm income (non-farm income includes income from non-agricultural economic activities like household and non-household manufacturing, handicrafts, repairs, construction, mining and quarrying, transport, trade, communication, community and personal services in the rural areas) in a span of 06 years since the announcement in 2016 requires a compound growth rate of 12.25% per annum. At the present real income growth rate of 5.24%, doubling would take around 14 years (Satyasai and Sandhya, 2016). Indian agricultural database lacks farmer income series and hence it becomes cumbersome to track the year to year changes in the household income. However, estimates from NSS data indicated that the compound annual growth rate (CAGR) of farmers' nominal income was 11.8% between 2002-03 and 2012-13 (Fig. 6.1). West Bengal had the lowest (6.7%) while Haryana registered the highest growth (17.5%). However, in real income terms, the growth was found to be lower relative to nominal income, wherein, Odisha topped with a CAGR of 8.3%. Livestock contribution resulted in a significant change for the state (Chandrasekhar and Mehrotra, 2016). It was closely followed by Haryana (8.0%), Rajasthan (7.9%) and Madhya Pradesh (7.3%), as against a national average of 3.5%. Bihar and West Bengal experienced negative growth (real terms) in terms of farmers' income. Under a Business as Usual (BAU) scenario, nominal income of a farmer can double almost every eight years, whereas doubling of real income needs more than 25 years. Clearly, if the aim is to double the real income by 2022, the effort and resources necessary to achieve it would have to be at least three times that of the current BAU-scenario levels (Gulati and Saini, 2016). In addition, survey data indicated that the farm income growth which is currently hovering around 1% has declined since 2011-12 (Chand *et al.*, 2015) and the growth in farm sector during April-June, 2017 quarter in nominal terms has been estimated as low as at 0.32% which is significantly lower for the first time in relative to real prices growth of 2.34% (Damodaran, 2017).

Table 6.1: Trends in Farmers' Income in India, 1993-94 to 2015-16

Year	Net value added at market prices (Rs. Crore)	Wage bills at market prices (Rs. Crore)	CPIAL (2004-05=100)	Total Farm income of all (Rs Crore)		Cultivars (Numbers in crores)	Farm Income per cultivator (Rs)	
				Market price	Real Prices		Current Price	Real Price
1993-94	223709	45755	59	177954	303814	14.39	12365	21110
1999-00	426582	90951	90	355631	372923	13.88	24188	26875
2004-05	527289	93130	100	434160	434160	16.61	26146	26146
2011-12	1409932	252804	183	1157128	632514	14.62	79137	43258
2012-13	1558480	245750	220	1312730	596695	14.36	91416	41553
2013-14	1753691	276532	245	1477159	602922	14.1	104763	42760
2014-15	1849931	291708	261	1558223	597020	13.85	112507	43106
2015-16	1940636	306010	273	1634625	598764	13.6	120193	44027

Source: Double Farmers Income National Institute of Transforming India GOI, New Delhi 2017

Notes: (1) Wage bills till the year 2011-12 are taken from Chand *et al.*, 2015. For subsequent years wage bills are estimated using the proportion of wage bill to net value added for the year 2012-13 (2) Farmers' income are expressed in real terms using CPIAL (2004-05=100) as deflator (3) Number of cultivars after the year 2011-12 were projected based on rate of growth between 2004-05 and 2011-12

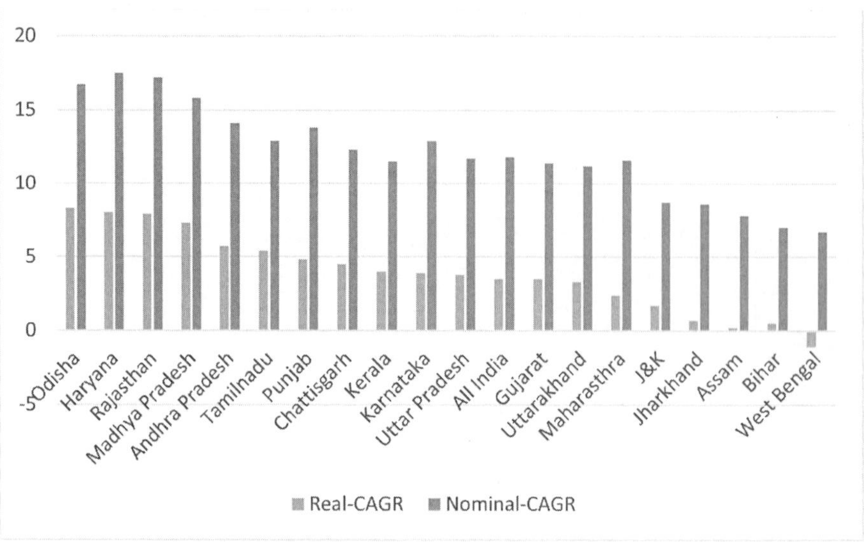

Fig. 6.1: Growth of Farmers Income between 2002-03 and 2012-13
Source: Gulati and Saini (2016)

Analysis of disaggregated sources of farmers' income indicates that for a majority of states, the real income and nominal income had doubled in the case of marginal farmers. Income from crop production, wages and livestock farming have increased across farm holding size against the declining non-farm income. Barring a few states like Arunachal Pradesh, Goa, Mizoram and Uttarakhand, the rest have experienced doubling in both real and nominal income between 2003 and 2013 (Fig. 6.2).

Income share from livestock has witnessed an increase from 5 to 12% at national level supporting the findings of Thiagu (2015) and across holding size, it increased from 6 to 14% in the case of marginal (<1 ha), 5 to 11% for small (1 to 2.5ha), 3 to 11% for semi-medium (2.6 to 4ha), 2 to 7% for medium (4.01 to 10ha) and 2 to 7% for large (> 10ha) land holders. The income growth attained across farm holdings and across regions shall be attributed to the livestock farming (Chandrasekhar and Mehrotra, 2016; Sendhil *et al.,* 2018)

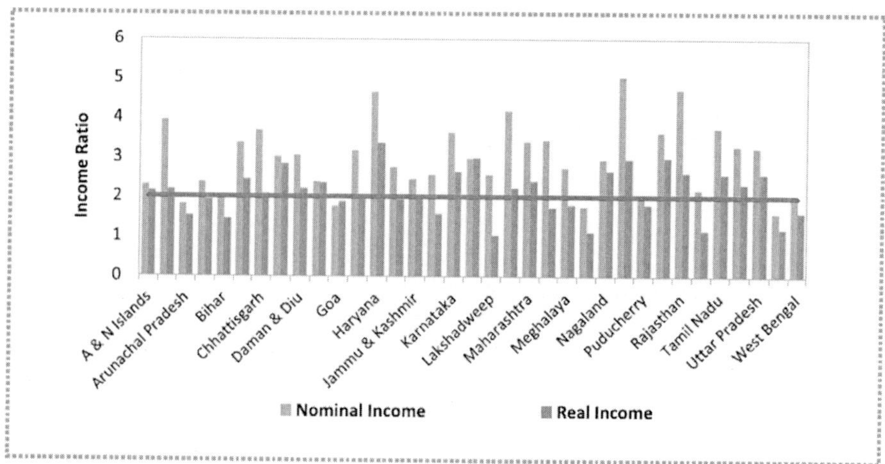

Fig. 6.2: Ratio of Farm household income between 2003-2013

6.3 Challenges Ahead in Agriculture

The challenges in agriculture are many in the context of doubling income. To cite a few, competition for land, water and energy; increasing cropping intensity particularly in the Indo-Gangetic plains leading to irrational resource use; changing pest complex; degradation of natural resources like land and water; declining total factor productivity; and stagnating yield (Sharma *et al.*, 2013). Indian agriculture not only faces the above routine challenges as it gets transformed but their intensity gets magnified *in lieu* of climate change. Agriculture not only affected by climate change but serve as a mitigator too. The production challenges are interdependent wielding the influence at different magnitudes across regions which finally gets reflected at macro level with negative impact. Hence, framing adaptation and mitigation strategies for climate change becomes utmost priority in Indian agriculture in the perspective of increasing the farm household income.

The complexity in addressing the production challenges embraced by the negative impact of climate change needs an in-depth understanding of the local situation for implementing relevant management strategies (Fig. 6.3). The implementation strategies to be successful at micro level need the synergy between the local communities and authorities apart from the support from government in terms of policies and investment on R&D to perform cutting-edge research (Chadha *et al.*, 2013). Further, price volatility (inputs and outputs) emerging from huge mismatch between supply and demand apart from risk and uncertainties in production have to be managed coherently adopting relevant farm practices (Sharma *et al.,* 2015) and/or price risk management tools (Sendhil *et al.*, 2013 & 2014a)

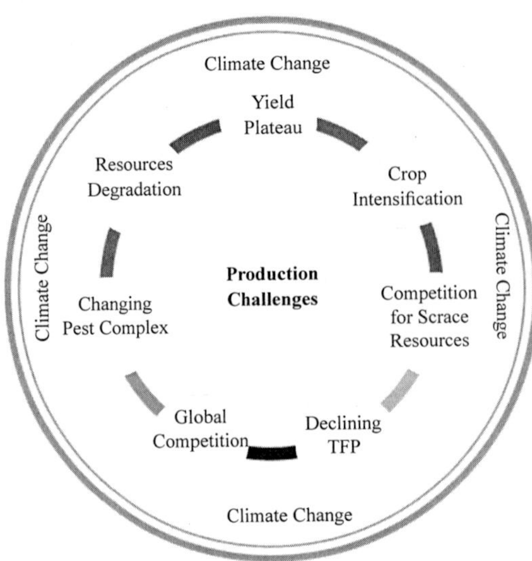

Fig. 6.3

6.4 Major Problems Faced by Indian Agriculture

6.4.1. Inequality in Land Distribution

The distribution of agricultural land in India has not been fairly distributed. Rather there is a considerable degree of concentration of land holding among the rich landlords, farmers and money lenders throughout the country. But the vast majority of small farmers own a very small and uneconomic size of holdings, resulting to higher cost per units. Moreover, a huge number of landless cultivators has been cultivating on the land owned by the absentee landlords, leading to lack of incentives on the part of these cultivators.

6.4.2. Land Tenure System

The land tenure system practiced in India is suffering from lot of defects. Insecurity of tenancy was a big problem for the tenants, particularly during the pre- independence period. Although the land tenure system has been improving during the post-independence period after the introduction of various land reforms measures but the problem of insecurity of tenancy and eviction still prevails to some extent due to the presence of absentee landlords and benami transfer of land in various states of the country.

6.4.3. Sub-division and Fragmentation of holdings

In India, the average size of holding is expected to decline from 1.5ha in 1990-91 to 1.3 ha in 2000-01. Thus the size of agricultural holding is quite uneconomic, small and fragmented. There is continuous sub-division and fragmentation of agricultural land due to increasing pressure of population and breakdown of the joint family system and also due to forced selling of land for meeting debt repayment obligations. Thus the size of operational holdings has been declining year by year leading to increase in the number of marginal and small holdings and fall in the number of medium and large holdings.

6.4.4. Cropping Pattern

The cropping pattern which shows the proportion of the area under different crops at a definite point of time is an important indicator of development and diversification of the sector. Food crops and non-food or cash crops arc the two types of crops produced by the agricultural sector of the country. As the prices of the cash crops are becoming more and more attractive therefore, more and more land have been diverted from the production of food crops into cash or commercial crops. This has been creating the problem of food crisis in the country. Thus after 50 years planning the country has failed to evolve a balanced cropping pattern leading to faulty agricultural planning and its poor implementation.

6.4.5. Instability and Fluctuations

Indian agriculture is continuously subjected to instability arising out of fluctuations in weather and gamble of monsoon. As a result, the production of food-grains and other crops fluctuates widely leading to continuous fluctuation of prices of agricultural crops. This has created the element of instability in the agricultural operation of the country.

6.4.6. Conditions of Agricultural Labourers

Agricultural labourers are the most exploited unorganized class in the rural population of the country. From the very beginning landlords and Zamindars exploited these labourers for their benefit and converted some of them as slaves or bonded labourers and forced to continue the system generation after generation. All these led to wretched condition and total deprivation of the rural masses. Now, the situation has improved marginally. But as they remain unorganized, thus economic exploitation of these workers continues. The level of income, the standard of living and the rate of wages remained abnormally low. Total number of agricultural workers has increased from 55.4 million in 1981 to 74.6 million in 1991 which constituted nearly 23.5% of the total working population of the country. This increasing number has been creating the problem of surplus labour or disguised unemployment, which in turn is pushing (heir wage rates below the subsistence level.

6.4.7. Poor Farming Techniques and Agricultural Practices

The farmers in India have been adopting orthodox and inefficient method and technique of cultivation. It is only in recent years that the Indian farmers have started to adopt improved implements like steel ploughs, seed drills, barrows, hoes etc. to a limited extent only. Most of the farmers were relying on centuries old. Wooden plough and other implements. Such adoption of traditional methods is responsible for low agricultural productivity in the country.

6.4.8. Inadequate Use of Inputs

Indian agriculture is suffering from inadequate use of inputs like fertilizers and HYV seeds. Indian farmers are not applying sufficient quantity of fertilizers on their lands and even the application of farm yard dung manure is also inadequate. Indian farmers are still applying seeds of indifferent quality. They have no sufficient financial ability to purchase good quality high yielding seeds. Moreover, the supply of HYV seeds is also minimum in the country.

6.4.9. Inadequate Irrigation Facilities

Indian agriculture is still suffering from lack of assumed and controlled water supply through artificial irrigation facilities. Thus the Indian farmers have to depend much upon rainfall which is neither regular nor even. Whatever irrigation potential that has been developed in our country, a very limited number of our farmers can avail the facilities. In spite of vigorous programme of major and minor irrigation projects undertaken since 1951, the proportion of irrigated land to total cropped area now comes to about 53% in 1998-99. Therefore, in the absence of assured and controlled water supply, the agricultural productivity in India is bound to be low.

6.4.10. Nonexistence of Crop Rotation

Proper rotation of crops is very much essential for successful agricultural operations as it helps to regain the fertility of the soil. Continuous production of cereals on the same plot of land reduces the fertility of the soil which may be restored if other crops like pulses, vegetables etc. are grown there. As the farmers are mostly illiterate, they are not very much conscious about the benefit of crop rotation. Therefore, land loses its fertility to a considerable extent.

6.4.11. Lack of Organized Agricultural Marketing

Indian farmers are facing the problem of low income from their marketable surplus crops in the absence of proper organized markets and adequate transportation facilities. Scattered and sub-divided holdings are also creating serious problem for marketing their products. Agricultural marketing in India is also facing the problem of marketing farmers' produce in the absence of adequate transportation and communication facilities, therefore, they fell into the clutches of middlemen for the speedy disposal of their crops at an uneconomic and cheaper price.

6.4.12. Instability in Agricultural Prices

Fluctuation in the prices of agricultural products poses a big threat to Indian agriculture. For the interest of the farmers, the Government should announce the policy of agricultural price support so as to contain a reasonable income from agricultural practices along with providing incentives for its expansion. Stabilization of prices is not only important for the growers but also for the consumers, exporters, agro-based industries etc. In India, the movements of prices of agricultural products are neither smooth nor uniform, leading a fluctuating trend. In the absence of proper price support and marketing support, prices of agricultural products has to go down beyond the reasonable limit so as to create a havoc on the financial conditions of the farmers. Again the exorbitant prices charged by the middlemen on agricultural crops also pose a serious threat to the consumers. Thus price, fluctuation may lead to disaster as both falling and rising prices of agricultural crops are having its harmful impact on the society as well as on the economy of the country.

6.4.13. Agricultural Indebtedness

One of the greatest problems of Indian agriculture is its growing indebtedness. The rural people are borrowing a heavy amount of loan regularly for meeting their requirements needed for production, consumption and also for meeting their social commitments. Thus the debt passes from generation to generation. Indian farmers fall into the debt trap as a result of crop failure, poor income arising out of low prices of crops, exorbitantly high rate of interest charged by the moneylenders, manipulation and use of loan accounts by the moneylenders and use of loan for various unproductive social purposes. Although they borrow every year but they are not in a position to repay their loans regularly as because either loans are larger or their agricultural production is not sufficient enough to repay their past debt. Thus the debt of farmers gradually increases leading to the problem of rural indebtedness in our country. Thus it is quite correct to observe that "Indian farmer is born in debt, lives in debt and dies in debt."

6.5 Challenges before the Modernization of the Indian Agriculture

The current phase of modernisation of Indian agriculture has thrown up some challenges, which cannot be ignored anymore. Unless these are squarely faced and comprehensive solutions worked out, the growth of the economy may be up against insurmountable barriers. The more important of these challenges are as follows:

6.5.1 Problem Relating to Rain-fed Crops

In respect of rain-fed crops such as coarse grains-poor men food-specially pulses, constraints on raising production are well-known. These are:

- An effective set of fully-developed technologies and extension methodology requires to be devised.
- High degree of uncertainty together with the relative poverty of the farmers makes the application of even known improved practices difficult and risky.
- The rural poor, particularly in the drought-prone areas and in remote areas of the country, continue to suffer from fluctuations in employment and income and inadequate availability of food grains in years of drought.

6.5.2 Problems relating to the Use of Farm Inputs

No less crucial to sustaining high growth rates in agriculture is the role played by farm inputs. Problems in this respect are the following:

- In regard to irrigation, through the area has shown a good improvement, the flow of benefit has not been commensurate. This is reflected both in the low intensity of cropping and in the underutilization of the potential created. The efficiency in the use of irrigation facilities also leaves much to be desired. In view of this, the productivity of irrigated land in the country is less than 50% of the potential.
- India has not been able to reach the targeted levels in the consumption of fertilizers. Even more important than the quantity consumed is the efficiency in the use of fertilisers. This has not been the case in the recent past, although there is a growing awareness of the problem. Besides this, the pattern of fertiliser consumption in the country is very highly skewed. In certain regions, a few crops and the rabi season account for the bulk of the fertiliser use.
- Recent plans have stressed the need for equitable and efficient distribution system, reduction in regional disparities and correction of the crop-wise imbalance that now exists in regard to various inputs.

6.5.3 Problems of Small Farmers

Over 80 million of 90 million operational farm holdings in the country are below 2 ha in size. About 60-70% of GDP from agriculture comes from subsistence agriculture. Unless small farmers are helped to improve their productivity and profitability through optimum use of their land, water, credit and other resources, it will not be possible to achieve our goal in food production of billion-plus. Decline in Productivity of Input: A major concern has been the decline in the productivity inputs. Various explanations have been given for declining productivity of inputs. These are:

- The new technology was initially adopted in areas with assured irrigation. The extension of new technology into more difficult terrains is bound to be more costly and capital intensive.
- Of late, there is evidence to suggest that new agricultural technologies are spreading in the rain-fed areas. The investment cost is substantially higher in the case of rain-fed areas than in the areas with assured irrigation.
- The fall in productivity could also have taken place because of the inefficiency in the use of inputs.

6.5.4 Rising Cost of Production in Agriculture

Over the last two decades, the prices received by the farmers have lagged behind the input prices, especially with regards to the prices of industrial inputs. The prices of important agricultural crops are reviewed every year by the government to keep them in line with costs of production and input prices. But often, what the farmers receive in effect is a weighted sum of the prices offered by the government and those prevailing in the free market. In this situation, the deteriorating terms of trade for farmers would mean either the inability of the government to compensate them inadequately for the increase in the cost of production and/or inadequate impact of the government measures on the free market prices.

6.5.5 Weakening of Linkages Between Agriculture and Industry

Although the inter-dependence of the agricultural and the industrial sectors has increased over the years, the strength of linkage between the agricultural and the manufacturing sectors has weakened. It must be realised that the majority of the population still depending on agriculture, and with a high rural bias in the culture of the population, weakening of this linkage is a serious matter of concern.

6.5.6 Highly Regulated Sector

Equally critical is the fact that agriculture is highly regulated that imposes restrictions on movement of agricultural products. Among these, the more important can be identified as:

- Compulsory levies (e.g. sugarcane, cotton).
- Licensing requirement (e.g. exports of agricultural products), and
- Internal trade restrictions (i.e. inter-state restrictions).

To sum up, although during the last six decades impressive gains have been made in agricultural production in some parts of the country, much still remains to be done to establish a balanced progress.

6.6 Some Important Outstanding Features of Indian Agriculture

6.6.1 Subsistence Agriculture

Most parts of India have subsistence agriculture. The farmer owns a small piece of land, grows crops with the help of his family members and consumes almost the entire farm produce with little surplus to sell in the market. This type of agriculture has been practised in India for the last several hundreds of years and still prevails in spite of the large scale changes in agricultural practices after Independence.

6.6.2 Pressure of Population on Agriculture

The population in India is increasing at a rapid pace and exerts heavy pressure on agriculture. Agriculture has to provide employment to a large section of work force and has to feed the teeming millions. While looking into the present need of food grains, we require an additional 12-15 Mha of land to cope with the increasing demands by 2010 A.D. Moreover, there is rising trend in urbanization. Over one-fourth of the Indian population lived in urban areas in 2001 and it is estimated that over one-third of the total population of India would be living in urban areas by 2010 A.D. This requires more land for urban settlements which will ultimately encroach upon agricultural land. It is now estimated that about 4 lakh hectares of farm land is now being diverted to non-agricultural uses each year.

6.6.3 Importance of Animals

Animal force has always played a significant role in agricultural operations such as ploughing, irrigation, threshing and transporting the agricultural products. Complete mechanization of Indian agriculture is still a distant goal and animals will continue to dominate the agricultural scene in India for several years to come.

6.6.4 Dependent upon Monsoon

Indian agriculture is mainly dependent upon monsoon which is uncertain, unreliable and irregular. In spite of the large scale expansion of irrigation facilities since Independence, only one-third of the cropped area is provided by perennial irrigation and the remaining two-third of the cropped area has to bear the brunt of the vagaries of the monsoons.

6.6.5 Variety of crops

India is a vast country with varied types of relief, climate and soil conditions. Therefore, there is a large variety of crops grown in India. Both the tropical and temperate crops are successfully grown in India. Very few countries in the world have a variety of crops comparable to that produced in India.

6.6.6 Predominance of food crops

Since Indian agriculture has to feed a large population, production of food crops is the first priority of the farmers almost everywhere in the country. More than two-thirds of the total cropped area is devoted to the cultivation of food crops. However, with the change in cropping pattern, the relative share of food crops came down from 76.7 percent in 1950-51 to 58.8 percent in 2002-03.

6.6.7 Insignificant place to given fodder crops

Although India has the largest population of livestock in the world, fodder crops are given a very insignificant place in our cropping pattern. Only four per cent of the reporting area is devoted to permanent pastures and other grazing lands. This is due to pressing demand of land for food crops. The result is that the domestic animals are not properly fed and their productivity is very low compared to international standards.

6.6.8 Seasonal pattern

India has three major crop seasons.

- **Kharif season** starts with the onset of monsoons and continues till the beginning of winter. Major crops of this season are rice, maize, jowar, bajra, cotton, sesamum, groundnut and pulses such as moong, urad, etc.
- **Rabi season** starts at the beginning of winter and continues till the end of winter or beginning of summer. Major crops of this season are wheat, barley, jowar, gram and oil seeds such as linseed, rape and mustard.
- **Zaid** is summer cropping season in which crops like rice, maize, groundnut, vegetables and fruits are grown. Now some varieties of pulses have been evolved which can be successfully grown in summer.

6.7 Doubling the Farmers Income

Agriculture continues to be the mainstay of Indian economy supporting 58% of Indian population, contributing 17.4% to the gross value addition with a growth rate of 4.1% during 2016-17. The sector has witnessed major quantum jumps in food production since Green Revolution and reckoned as a surplus nation in food grains production. Despite a leader in producing several agricultural commodities, farmers' income across regions and holding size has not posted the desired level of growth. The average per capita income of farmers is just one-fifth of the national average. Farmers face the risky and uncertain situation nearly every day in their farm business which aggravates their stress on continuation of agriculture as a profession posing a severe threat on their livelihood and sustainability.

In 2016, Prime Minister Modi first declared his government would double farm incomes in other words, raise them by 100% in six years by 2022-23, a politically significant goal in a country where nearly half the population depends on a farm-based livelihood. Prime Minister Narendra Modi had shared his vision of doubling farmers' income for the first time while addressing a Kisan Rally at Bareilly in Uttar Pradesh on February 28, 2016. Since then, the Central government has been giving more thrust to the growth of agriculture and allied sectors.

The agriculture ministry has drawn up a 7-point strategy to double farmers rsquo; income in the country.

1. Big focus on irrigation with large budgets, with the aim of per drop, more crop.
2. Provision of quality seeds and nutrients based on soil health of each field.
3. Large investments in warehousing and cold chains to prevent post-harvest crop losses.
4. Promotion of value addition through food processing.
5. Creation of a national farm market, removing distortions and e-platform across 585 stations.
6. Introduction of a new crop insurance scheme to mitigate risks at affordable cost.
7. Promotion of ancillary activities like poultry, beekeeping and fisheries.

Agricultural strategy in the country during the planned development era has been to ensure food security and farmers have responded to the nation's needs well and adopted Green Revolution technology. While the country achieved commendable position in food production, farming itself turned non-profitable overtime due to rising costs and uneconomical holdings. The spreading of harmful ideas of farmers' distress across the country has shaken the agrarian foundations. Enhancing incomes of the farmers and ensuring their income security, thus, has been of concern to all. Unless farmers' income increases substantially, distress cannot be tackled (Chand, 2016). National Commission on Farmers under the chairmanship of Dr .M. S. Swaminathan has addressed the issue of distress and farmers' welfare through a series of recommendations. The Hon'ble Prime Minister in an address to farmers in District Sheopur in Madhya Pradesh exhorted to double the incomes of farmers by 2022. Subsequently, the announcement was formalized in the Union Budget 2016-17 stating that an important objective of the Government is to double the income of farmers by the year 2022. Trend in Farmers' income: All India There are hardly any data sources that can give income estimates for farmers. We have earlier efforts by Dholakia *et al.,* (2014) and Chand et al (2015). As per the Chand *et al.,* 2016 estimate, the real income grew at the compounded rate of 3.94% per annum during 2004-05 to 2011-12 which is the fastest compared to previous two decades. Based on the trends in farm income from 1983-84 till 2011-12, Chand *et al.,* (2015) concluded that:

- The income earned by farmers net of input cost and wage bill has seen low and high growth paths in different periods.
- The growth in farm income accelerated towards recent period ending 2011-12.
- Decent growth in farm income requires high growth in output, favorable farm produce prices and some cultivators moving out of agriculture.
- A high growth in agriculture can reduce income disparities and promote inclusive growth.
- Low growth of farm income seems to have been associated with agrarian distress, and,
- More than half of farm households in the country would remain below poverty level unless they adopt high-income earning avenues and augmenting their incomes through non-farm activities.

The research bulletin brings out the rationale for doubling the farmers' income, followed by tracking the income status of farm households along with pathways and strategies for attaining the income doubling by 2022. Income is the most relevant measure to assess farmers' economic well-being and sectoral transformation. The crises and distresses plaguing the sector endanger the very livelihoods and welfare of the farmers. Indian Government with the intention giving enough policy thrust on income security, proposed to double the farmers' income by 2022, platinum jubilee year of the India independence. The present study analysed the current status of farmers' income across holding size and regions and attempted to decipher the scope and pathways for doubling income through potential drivers. The spatial and temporal trends in farm household income from crop production, livestock farming, wages and non-farm activities have been analysed for better understanding of the present scenario. The study tracked the farmers' household income across regions and holding size between 2003 and 2013 using the countrywide situation assessment survey data collected by the National Sample Survey Office (NSSO) and deciphered the scope and potential for doubling the farmers' income (DFI) by 2022. The findings from the NSSO data indicated that the share of income has increased drastically from 5 to 12% in the case of livestock farming, 45 to 48% in crop production, while that of the wages and non-farm have declined between 2003 and 2013. The challenges faced by the farming community in the coming years have been highlighted for devising relevant pathway and strategies to enhance the income.

Yield enhancement followed by cost reduction, fair price realization and risk adaption has been identified as the potential pathway for doubling income. Farmers' income from crop production, livestock farming, wages and non-farm activities is an outcome of synergy and convergence between technology, extension, institutions and policies to achieve the set target. Indian agriculture needs a relook with a special focus on farm income through productivity/efficiency enhancement coupled with cost reduction, better price realization and income risk coverage to be on the track of DFI by 2022. The study also furnished strategies for doubling the income of wheat/barley producers focusing on varietal adoption, resource conservation technologies, diversification / intensification / relay cropping, pest and disease management in vulnerable regions, and cost reducing technologies. The study concluded that the government has to integrate investment and leadership in science & technology, extension, institutions and policy interventions to accomplish the set goal by 2022.

Farmers' income is directly related to cost of agricultural production (including input costs) and profitable monetisation of the agricultural produce, through effective market linkages. The DFI Committee Report, in Volumes III – XI and XIII, deliberates upon specific economic activities and topics that have a durable impact on farmers' income increase, some of which are categorised as follows:

6.7.1 Demand Driven Agricultural Logistics System for post-production operations such as produce aggregation, transportation, warehousing, etc.

- **Agricultural Value System** (AVS) as an integration of the supply chain and to drive market led value system – District level, State level and National Level Value-System Platforms to promote individual value chains to collaborate and integrate into a sector-wide supply chain.
- **Farmer- Centric National Agricultural Marketing System** by restructuring of the marketing architecture and networking of Primary Rural Agriculture Markets (22,000) and wholesale markets to facilitate pan-India market access; as also integrating the domestic market with export market.
- **Developing Hub and Spoke System** at back-end as well as front-end to facilitate and promote the AVCS (which includes input providers, farmers, transporters, warehousing, food and agro-processors, retailers).
- **Marketing Intelligence System** to provide demand led decision making support system - Forecasting system for agricultural produce, supply and demand, and crop area estimation to aid price stabilisation and risk management.
- **Promoting Sustainable Agriculture** – Conservation Agriculture, Climate Resilient Agriculture, Rainfed Agriculture, Ecology Farming, Watershed Management System, Integrated Farming System, Agro-Climatic Regional Planning, Agricultural Resources Management and Micro-Level Planning, etc.
- **Effective Input Management** achieving Resource Use Efficiency (RUE) and Total Factor Productivity (TFP) – Water, Soil, Fertilisers, Seeds, Labour-Farm mechanization, Credit and Precision farming, so as to reduce farm losses.
- **Enhancing Production through Productivity** – to achieve & sustain higher production out of less and release land and water resources to diversify into higher value farming for enhanced income.
- **Farm Linked Activities** including secondary and tertiary sector activities of MSME scale, for promoting near-farm and off-farm income generating opportunities as well as to facilitate that more of the produce captures more of the market value.
- **Agricultural Risk Assessment and Management** including drought management, demand & price forecast, weather forecast, management of biotic stress including vertebrate pests access to credit for farmers

for farming operations; providing long term credit, post-production finance to preventing distress sale by farmers, and crop & animal risk management through insurance.

- **Empowering Farmers** through Agricultural Extension, Knowledge Diffusion and Skill Development.
- **Research & Development and ICT** for Doubling Farmers' Income.
- **Structural and Governance Reforms in Agriculture,** including building a database of farmers, facilitating farmer & produce mobilisation, institutional mechanism at district, state & national levels for coordination & convergence, utilising Panchayat Raj Institutions as key delivery channels for transparent and inclusive development.

6.8 Pathway for Doubling the Farmers Income

The scope for DFI by 2022 exists with the following options viz., increasing the physical output from agriculture, diversification of enterprises, pricing mechanism, adoption of risk management tools, wage rate and salaries for farm labour as well as different sources of non-farm income (NITI Aayog, 2015; Chand, 2017; Saxena et al., 2017). A pathway has been formulated which is applicable to all commodities and sectors in agriculture (Sendhil et al., 2017a & 2017b). However, the implementation strategies should be region and need specific targeting different components in agriculture. The framework integrating technology, extension, institutions and policies to double the farmers' income has been depicted in Fig. 6.4.

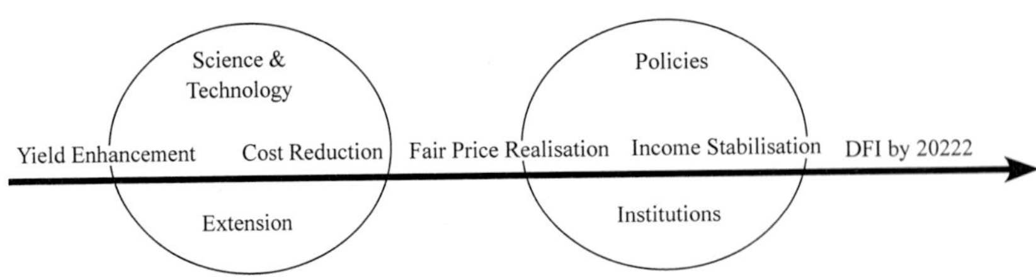

Fig. 6.4: Pathway for doubling the farmers; Income (DFI) by 2022

6.7.1 Technology

Technology - the outcome of science - enables increased output with the same input or realise the same output with reduced input. Productivity in a majority of the commodities has struck a plateau demanding barrier breaking intervention through cutting-edge sciences. The diversification of farm activities towards high value crops and enterprises can more than quadruple income from the same piece of land. Another augmenting factor for doubling farm income is through integrating crop production with allied and subsidiary enterprises in uncultivated lands.

6.8.2 Extension

Given the technology and resources, the output level can be enhanced by consolidating the existing potentials by bridging the yield gaps between agronomic potentials achieved in research and extension farms and the actual yields obtained in average farmer's fields. Around 30 per cent of the physical output can be increased which will facilitate to increase the farmers' income without any additional investment, but just by adopting the recommended package of practices across different sub-sectors in agriculture.

6.8.3 Institutions

Irrigation is the best insurance against drought. The thrust of the state is to ensure 'Per drop more crop' through accelerated irrigation schemes supplemented by massive promotion of micro irrigation techniques for maximum coverage of irrigated crop area. As agriculture in India is a gamble with the monsoon, and the economic pursuit of the enterprise is worth if and only when adequate risk management options are available. The recent initiatives of the government that is likely to impact agriculture can be broadly classified as those can prevent the volatility in farm income and those can improve the farm productivity. Any amount of productivity increase unless accompanied by farm income increase either by direct price incentive or cost reduction or economy of scale or value addition or anything else will have no meaning for farmers. Ironically high production is always associated with low prices and vice versa. Hence, there is a need for innovative marketing that will link farmers to productivity tools and insurance. The Pradhan Mantri Krishi Sinchai Yojana (PMKSY) promotes more crop per drop as a national mission to improve farm productivity with a focus on massive expansion of micro irrigation at farm level and by closely monitoring 99 priority projects with a potential of irrigating over 7.6 million ha for a faster completion. These will reduce the monsoon dependency insulating from shocks and improve productivity.

Prime Minister's Crop Insurance Scheme, is based on crop income than cost of cultivation/credit and thus addresses volatility in farm income besides natural calamities and post-harvest losses unlike earlier schemes that covered only loss of yield or investment or capital besides. It is planned to enhance the coverage of the scheme to more than 50% of the cropped area by next year. This coupled with land leasing laws will enable farmers including tenants to absorb risk and invest more into better farm techniques and crops. The recently initiated National Agricultural Marketing brings more than 500 markets on a single e-platform integrating markets sans middlemen and enabling farmers to bid their products to sell anywhere in India. But, the integration will have operational meaning for the farmers only when Agricultural Produce Market Committee (APMC) Act and Essential Commodities Act (ECA) are either revamped or done away with. This is very much evident in the difference in realization of the price potential between livestock products not covered under these acts and the fruits and vegetables covered under the APMC. Further, it will not suffice if commodities are delisted under APMC unless alternate marketing platforms are made available with attendant, scientific storage, processing and transport facilities where private can play a vital role.

6.8.4 Policy

Policy on agricultural commodities export and stocking have more often than not acted against the farmers' interest preventing them from par taking the benefits of free market. Whenever prices shoot up, banning or restricting export or limiting stocking of the commodity have been resorted to more as a panic reaction than a planned strategic measure. By these measures, India has gained notoriety for being an unreliable supplier for international market denying its farmers their due benefits from participating in the international trade. Besides, the uncertainties regarding stocking norms, no investment worth its while can ever happen on scientific storage, processing, value addition and contract farming. Unless agricultural commodities are freed from the ambit of Essential Commodities Act or the Act is repealed altogether, income from and investment on agricultural marketing is not possible. When there is a spike in prices, the difference between modal market prices and MSP may be made up through direct transfer of deficiency payment to farmer.

Terms of trade is another important policy tool to enhance farmers' income by tweaking the prices of farm products compared to their non-agricultural counterparts but without stoking food inflation? Inflation in agricultural prices also leads to an enhanced real farm income if prices received by farmers increase at a faster rate in comparison to the prices paid by them. For the past few years, the wholesale price index or WPI based inflation of non-agricultural prices is declining, whereas that of the agricultural prices has been increasing by about 5 per cent (in the year 2015-16) implying a 5% growth in real farm income. If technology and factor prices could result in per-unit cost savings, farmers' income would rise at a much higher rate than the rate of increase in output.

6.9 Strategies for Doubling Income

Income is the most relevant measure to assess the farmers' welfare and agriculture transformation. Even today, the highest returns on investment on per unit basis are from agriculture. What is lacking is the scale unlike corporate investment. Certainly, returns from cultivation alone will not help to achieve the set target of DFI. It has to be supplemented, in fact to a larger extent by livestock and other non-farm activities supported with policy intervention at all levels (Table 6.2 and 6.3) (Chand, 2017).

Table 6.2: Potential strategies to enhance annual farm household income

Parameters	Strategies
Science & Technology	• Adoption of improved varieties/breeds/strains for additional income • e-Grid of all weather stations for providing location specific weather information • Nutrients sale based on Soil health card programme • Adoption of micro irrigation system to improve the water use efficiency
Extension	• Bridging the gaps between achievable (FLD) and potential yields • More number of effective cluster demonstrations to bridge the information gap • Village adoption to transfer the technologies developed by the research organizations • Upscaling and out scaling of technologies through field days, exhibition and other activities
Policies	• Rationalizing the subsidy on energy use. • Enrolling more number of marginal and small holders under crop insurance scheme. • Integrating all central and state subsidies in agriculture • Formation of Crop Planning Department at national and state level. • Additional investment on agricultural R&D to pave path for innovation. • Setting up more organic food certification agencies • Policy for setting up of FPO for block level seed production. • Integrated land-use policy particularly for water • Breaking of crop monotony and focus on diversification
Institutions	• Setting up of Agribusiness Centres at district level • Transparency and simplified procedures in electronic trading • Developing comprehensive framework for community / corporate farming. • More emphasis on e-learning in regional languages. • Value chain development for primary commodities

Table 6.3: Potential strategies to enhance income across sources Parameter Strategies

Parameter	Strategies
Crop	• Expansion in irrigation coverage • Improvement in quality seed production and distribution • Improvement in market connectivity • Sustaining/increasing the price support
Livestock	• Promoting value-addition at the household level • Increasing the area under fodder
Wages	• Skill training in agricultural operations • Cost-effective mechanization
Non-farm	• Rural industrialization • Skilling in different non-farm activities

Ministries / Departments of Government of India, representing the panoply (range) of the complexities that impact the agricultural system. Members drawn from the civil society with interest in agriculture and concern for the farmers were appointed by the Government as non-official members. The DFI Committee has co-opted more than 100 resource persons from across the country to help it in drafting the report. These members hail from the world of research, academics, non-governmental organisations, farmers' organisations, professional associations, trade, industry, commerce, consultancy bodies, policy makers at central & state levels and many more of various domain strengths. Such a vast canvas as expected has brought in a kaleidoscope of knowledge, information, wisdom, experience, analysis and unconventionality to the treatment of the subject. The Committee over the last more than a year since its constitution vide Government O.M. No. 15-3/2016-FW dated 13th April, 2016 has held countless number of internal meetings, multiple stakeholder meetings, several conferences & workshops across the country and benefitted from many such deliberations organised by others, as also field visits. The call of the Hon'ble Prime Minister to double farmers' income has generated so much of positive buzz around the subject that no day goes without someone calling on to make a presentation and share views on income doubling strategy. The Committee has been, therefore, lucky to be fed pro-bono service and advice. To help collage, analyse and interpret such a cornucopia of inputs, the Committee has adopted three institutes, namely, NIAP, NCAER and NCCD. The Committee recognizes the services of all these individuals, institutions & organisations and places on record their service. The Committee has adopted a basic equation of economics to draw up its strategy, which says that net return is a function of gross return minus the cost of production. This throws up three variables, namely, productivity gains, reduction in cost of cultivation and remunerative price, on which the committee has worked its strategy. In doing so, it has drawn lessons from the past and been influenced by the challenges of the present & the future. In consequence, the strategy platform is built by the following four concerns:

a) Sustainability of production

b) Monetisation of farmers' produce

c) Re-strengthening of extension services

d) Recognising agriculture as an enterprise and enabling it to operate as such, by addressing various structural weaknesses.

The Doubling Farmers' Income (DFI) Committee recognises agriculture as a value led enterprise and suggests empowering farmers with "improved market linkages" and enabling "self-sustainable models" as the basis for continued income growth for farmers. This builds the basic strategy direction for four primary concerns: optimal monetization of farmers' produce, sustainability of production, improved resource use efficiency and re-strengthening of extension and knowledge based services. The Committee identifies and focuses on seven major sources of growth (Volume II), operating within and outside the agriculture sector. These are,

a) Improvement in crop productivity.

b) Improvement in livestock productivity.

c) Resource use efficiency or saving in cost of production.

d) Increase in cropping intensity.

e) Diversification towards high value crops.

f) Improvement in real prices received by farmers.

g) Shift from farm to non-farm occupations.

6.10 Digital Technology in Agriculture

Agriculture is the world's first solar powered factory. It organizes human effort to capture the life force hidden in seed and sunlight, to process the elements in the soil, air and water, into produce. Agriculture converts hard natural elements into useful material for use of mankind. Agriculture, in itself, is not digital but a very material physical process. Like in all other parts of one's life, there is increasing use of electronic devices, tools and a fusion of digitised systems, to manage agricultural processes. Hence, the phrase "**Digital Agriculture**", a recently coined catch-phrase seems a misnomer - it actually refers to the use of digital technologies in managing the business of agriculture. '**Digital Technologies in Agriculture**' is probably the more accurate narrative.

Advanced technologies are not uncommon to agriculture. Gene mapping under the electron microscope, bio-technology and the information services have regularly brought into use as available digital technologies. Satellites &

weather scanning radars and digital temperature & humidity sensors have been around for a few decades. However, of late, the digitization of agricultural information and its analytics is increasingly changing the way farming is done. Gone are the days of second guessing the next monsoon, or of watering the field at set schedules. Embedded sensors in the soil can tell when and how much water to feed, and digital analysis of wind patterns and upper jet streams has reduced the guess work of rain dancers. Digital photography allows for near-instant spectrographic analysis of soil and plant health. In minutes, patterns not visible to the eye, are compared with a large database to diagnose the INM and IPM needs on a farm. These possibilities are already in use in pockets, and need to find more universal application.

Digitization of everyday data, helps to analyse trends, forecast events in advance and quickly share knowledge across geographies. Already, in the communications arena, digital technologies have made a huge impact on agricultural communities. Alerts and remote advices are frequently sent to farm across the digital ether helping in intelligent decision making. Digital technologies can help counter many of the inherent vagaries in farming and optimise upon the resources that are deployed. Given, that there is increasing concern from climate change, land depletion, water shortage and wasteful use of agro-chemicals, making sure that farms remain environmentally relevant and sustainable will require more use of digital devices and the associated analytics in agriculture.

A farm itself is no longer a large field, but can now be addressed by aid of technology at the scale of a few feet and even at a granular level. Geo-informatics is not merely the cartography of large tracts of land, but can be used for precisely targeted treatment on a field, without affecting the adjoining tract of land. This precision can even happen at the level of individual plants, opening up opportunities for efficiency, hitherto unimaginable. While one is used to seeing growing use of protected cultivation, using physical plastic and glass to envelope and safeguard the plants from inclement happenings, the future may yet see another envelope, invisible and digital, that is superior to the current day protected cultivation practices.

In agriculture linked governance, a fusion of digital technologies are playing an important role. The direct benefit transfer (DBT) system for a multitude of support schemes, linked with Aadhaar (the 12 digit unique identity number based on biometrics) is a notable example. Digital head count of livestock and analysis of their health through RFID (radio frequency identification) and micro-chips based ear tags and the traceability of a vegetable to its farm plot through digital barcodes are examples of applications. The electronic National Agricultural Market (eNAM) is yet another example, where digital technologies are aimed to link farmers seamlessly with a national level market.

Secondary agricultural activities, involving the post-production management phase of produce, is another area seeing increasing use of digital technologies. Big data analytics that can forecast consumer demand, not just a week in advance, but even before the farmer plans to sow their field, will redefine crop planning, harvest scheduling and market linkages. There is need to ensure that the output from farms will find a market, across place, time and form. Digital data collation and analytics will be a big boon in this area, especially in India where though producing surpluses, yet many still remain hungry. Not too far into the future, any citizen could order a flower or a fruit while still on the bush, thanks to use of digital technologies in agriculture. Emergent Artificial Intelligence (AI) may replace the shepherd dog, or take self-contained decisions to robotically prune leaves and manage the canopy in orchards. Maybe, farmers will have a robot repair workshop, with 3-D printers in their barns, in a few more decades.

The application of modern day technologies, especially digital technologies are all pervading, and this Committee does not intend to list all the options and applications. The options available include innovations, which also change with every new development. However, there is a need to prioritise efforts, and the localised need assessment of farming clusters at state and district level, and should be fully considered. This will make the S&T system more effective and efficient.

At the national level, the priority areas to target doubling of farmers' income, though Science and Technology (Fig. 6.5) could be:

- **Famers database** – as recommended in Volume-XIII, to build a dynamic database and ensure targeted and efficient delivery of support to farmers, and to assist specialised extension services.
- **Credit availability** – to provide greater coverage under Kisan Credit Cards including crops, fishers and livestock farmers, and universal access to post-harvest pledge loans.
- **Market efficiency** – to provide market intelligence through demand & price forecasting.
- **Extension system** – to standardize the information, integration of effort among stakeholders and to maximise coverage to reach all farmers.
- **Resource use efficiency** – specifically to improve soil and water management.
- **Sustainability and productivity gains** – to improve yields and broad base the production while suiting regional ecological strengths.

- **Risk management** – information and insurance systems that improve farmers' capacity to handle the vagaries of weather, pests, disaster and markets,
- Convergence in efforts by public and private sectors.

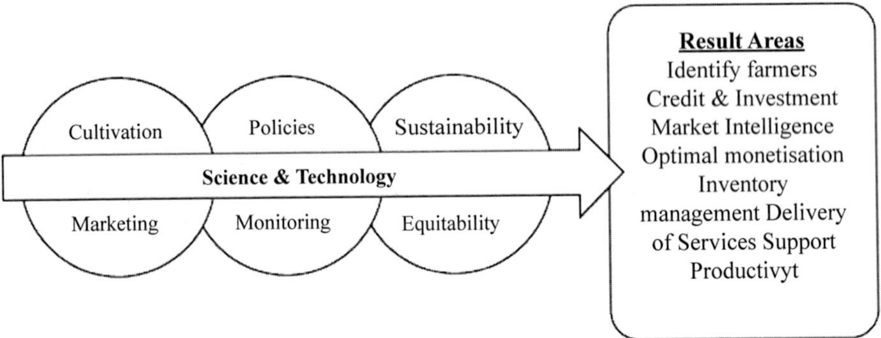

Fig 6.5: Priority Areas to target doubling of farmers' income, though Science and Technology at National level

There are various schemes and programmes of both Public Sector (Ministry of Agriculture & Farmers' Welfare, Ministry of Water Resources, Ministry of Rural Development, Ministry of Fertilizers, Ministry of Forests, Environment and Climate Change, Ministry of Science & Technology, Department of Space, Ministry of Earth Sciences, Ministry of Electronics and Information Technology, Ministry of Communications, Ministry of Commerce and Industry, Ministry of AYUSH, Ministry of Micro, Small and Medium Industries etc.) and Private Sector at village level, and digital technologies that can help bring convergence at ground level (DFI, 2018).

Farmers' income is directly related to cost of agricultural production (including input costs) and profitable monetisation of the agricultural produce, through effective market linkages. The DFI Committee Report, in Volumes III – XI and XIII, deliberates upon specific economic activities and topics that have a durable impact on farmers' income increase, some of which are categorised as follows:

a) **Demand Driven Agricultural Logistics System** for post-production operations such as produce aggregation, transportation, warehousing, etc.

b) **Agricultural Value System** (AVS) as an integration of the supply chain and to drive market led value system – District level, State level and National Level Value-System Platforms to promote individual value chains to collaborate and integrate into a sector-wide supply chain.

c) **Farmer - Centric National Agricultural Marketing System** by restructuring of the marketing architecture and networking of Primary Rural Agriculture Markets (22,000) and wholesale markets to facilitate pan-India market access; as also integrating the domestic market with export market.

d) **Developing Hub and Spoke System** at back-end as well as front-end to facilitate and promote AVCS (which includes input providers, farmers, transporters, warehousing, food and agro-processors, retailers).

e) **Marketing Intelligence System** to provide demand led decision making support system - Forecasting system for agricultural produce, supply and demand, and crop area estimation to aid price stabilisation and risk management.

f) **Promoting Sustainable Agriculture** – Conservation Agriculture, Climate Resilient Agriculture, Rainfed Agriculture, Ecology Farming, Watershed Management System, Integrated Farming System, Agro-Climatic Regional Planning, Agricultural Resources Management and Micro-Level Planning, etc.

g) **Effective Input Management** achieving Resource-Use-Efficiency (RUE) and Total Factor Productivity (TFP) – Water, Soil, Fertilisers, Seeds, Labour-Farm mechanization, Credit and Precision farming, so as to reduce farm losses.

h) **Enhancing Production through Productivity** – to achieve & sustain higher production out of less and release land and water resources to diversify into higher value farming for enhanced income.

i) **Farm Linked Activities** including secondary and tertiary sector activities of MSME scale, for promoting near-farm and off-farm income generating opportunities as well as to facilitate that more of the produce captures more of the market value.

j) **Agricultural Risk Assessment and Management** including drought management, demand & price forecast, weather forecast, management of biotic stress including vertebrate pests access to credit for farmers for farming operations; providing long term credit, post-production finance to preventing distress sale by farmers, and crop & animal risk management through insurance.

k) **Empowering Farmers** through Agricultural Extension, Knowledge Diffusion and Skill Development.

l) **Research & Development and ICT** for Doubling Farmers' Income.

m) **Structural and Governance Reforms in Agriculture,** including building a database of farmers, facilitating farmer & produce mobilisation, institutional mechanism at district, state & national levels for coordination & convergence, utilising Panchayat Raj Institutions as key delivery channels for transparent and inclusive development.

6.11 Role of Research & Development (R&D) in Indian Agriculture

Agriculture is the backbone of the country with about 48 per cent of the population dependent upon it and more than 65 per cent of the citizens living in rural India. It is the source of food, feed, fibre, fuel and the livelihood of Indian people. Increasing agricultural productivity is a key challenge for in realising higher output and farmers' income. The green revolution endowed India with a greater genetic diversity, and supported by enhanced institutional capacity, led to produce more crops. The introduction of high yielding varieties, additional irrigation facilities, a great input flow through fertilizers and pesticides, farm mechanization, credit facilities, buttressed by price support, and other rural infrastructure facilities ushered in the green revolution. It stimulated infrastructure and rural development, increased prosperity of villages, and improved the quality of life. The radical change in land use and agricultural production transcended India from a food importing country to a self-sufficient and even to a food-exporting nation. There is lot of improvement in the agricultural production and productivity *per se* in India after the green revolution (Table 6.4).

Table 6.4: Increase in productivity (kg/ha) of food grains in India

Year	Foodgrain production (million tonnes)	Foodgrain productivity (kg/ha)
1960-61	82.02	710
1970-71	108.42	872
1980-81	129.59	1023
1990-91	176.39	1380
2000-01	196.81	1626
2010-11	244.49	1930
2015-16	251.57	2042
2016-17	275.11	2129

6.12 Status of Research & Development

The country has witnessed an impressive growth in rice production in the post-independence era due to the adoption of semi dwarf high yielding varieties coupled with the adoption of intensive input based management practices. Rice production has increased four times, productivity three times while the area increase has been only one and half times during this period. In order to keep pace with the growing population, the estimated rice requirement by 2025 is about 130 million tonnes. Plateauing trend in the yield of HYVs, declining and degrading natural resources like land and water and acute shortage of labour make the task of increasing rice production quite challenging. The current situation necessitates adoption of some innovative technologies to boost rice production. The country has become a leading exporter of the Basmati rice. Major export destinations (2016-17) were Saudi Arabia, Iran, United Arab Emirates, Iraq and Kuwait. The country exported 40,00,471.56 metric tons (Mts) of Basmati rice worth of Rs. 21,604.58 crore (or 3,230.24 US$ Mill.) during the year 2016-17 to these countries. Other area of work to enhance yield is popularization of rice hybrid. Efforts are being made to promote cultivation of hybrid rice through various crop development programmes such as National Food Security Mission (NFSM), Bringing Green Revolution to Eastern India (BGREI) under Rashtriya Krishi Vikas Yojana (RKVY). From the initial level of 10,000 ha in 1995, area under hybrid reached one m ha in 2006, exceeded 2.5 mha during 2014, which is about 5.6 per cent of the total rice area in the country. It has picked up during the last eleven years, mainly because of increasing popularity of hybrid rice in Uttar Pradesh, Bihar, Jharkhand, Madhya Pradesh and Chhattisgarh and it is estimated that around 3 m ha plus was under hybrid rice cultivation in India in 2016 which is around 7 per cent of the total rice cropped area in India. Maize (corn) is one of the most versatile emerging crops having wider adaptability under varied agro-climatic conditions and is valued as food, feed, fodder and industrial raw material. Globally, maize is known as the queen of cereals, because it has the highest genetic yield potential. In India, maize is the third most important food crop after rice and wheat. In view of its universal nature and adaptability under diverse ecology, development of high yielding hybrids with in-built resistance and tolerance to diseases, pests and various climatic stresses; and development and fine-tuning of production ecology deserve priority attention. During the last 10 years, more than 120 hybrids have been developed and released in addition to development of various production technologies. As of now (2017-18) more than 200 cultivators are available in the field. While it is true that the growth rate of maize output over the last 5 years is more than that of USA & China, the country's output of 27 million tonnes (2016-17) pales before the output of 370 million tonnes of USA. Productivity in India is 2.6 tonnes/ha as against 12 t/ha in USA and 5.5 t/ha in China.

Considerable field scale understanding is available in India from empirical research on crop response to irrigation under varying soil, water use and environmental conditions, and efficient irrigation schedules have been developed for all the major crops. Similarly, technology improvements are being consistently addressed to application efficiencies of the commonly practised basin, flood, furrow and border methods of irrigation. Similarly, newer technologies are becoming increasingly available to help reduce the energy requirements of pressurized micro-irrigation systems like drip and sprinkler. Low pressure micro-irrigation technologies are being aggressively promoted in India by the central government, state governments and many non-governmental organisations (NGOs), to improve irrigation efficiencies and agricultural productivity.

The new technologies of micro-irrigation now include drip/trickle systems, surface and subsurface drip tapes, micro-sprinklers, sprayers, micro- jets, spinners, rotors, bubblers, etc. Despite wide promotion, only about 10 million ha of land is currently under micro-irrigation in India as against the total potential of 63 million ha. Maharashtra is the leading state under micro-irrigation followed by Karnataka, Andhra Pradesh and Tamil Nadu. Micro irrigation is popularly practised in about 30 crops, and is more popular in horticultural crops which allow relatively wide spacing. It is however critical that micro-irrigation is popularised and facilitated in field crops grown in rainfed cultivation systems. This will benefit the small and marginal farmers, who are predominant practitioners of field crops and rainfed farming systems.

Studies have revealed that water savings ranging between 25 and 50 percent are possible by drip irrigation compared with surface irrigation. Micro-irrigation also facilitates application of controlled quantity of water and nutrients in the vicinity of each plant, such that the crop, water and nutrient needs are almost matched with irrigation water supplies. Most current research in micro-irrigation is focused on simultaneous precision application of water, fertilizer and other inputs to match the crop requirements on the field to increase the marginal productivities of water and inputs through impact on the quantity and quality of produce.

The practice of conservation tillage, agro-forestry, ley (fodder/pasture) farming, mulching, nitrogen fixing legumes, crop diversification, integrated nutrient management etc has the potentiality to enhance C-sequestration in various locations.

Freshwater aquaculture is an integral part of the agriculture in India. It is one of the fastest growing subsectors in the country which has registered a growth rate of 5.1 percent per annum over the last 60 years. During this period, the fish production in the country has increased from 0.75 million metric tonnes (mMT) in 1950-51 to 9.45 mMT in 2013-14, of which the major contribution has been from aquaculture as the sector has grown from 0.37 mMT in 1980 to 5.1 mMT at present. The consumption demand for fish is rising over a period of time, primarily due to the growing population, expanding urbanization and changing food habits. In future, freshwater aquaculture sector holds the key as around 85 percent of the additional food fish demand could be met from the freshwater sector. The dairy and livestock sector too has grown and both milk and meat production have registered impressive growth, and India tops the global milk output at 165 million tonnes.

6.13 Issues and Challenges

The national agricultural research and development system aided by several organisations and institutes has taken up many activities to develop and demonstrate new technologies for strengthening agricultural farming including dairy, livestock and fisheries in the country. The science and technology-led development in agricultural farming has resulted in manifold enhancement in productivity and production of different crops and commodities to match the pace of growth in demand. But, at present the agricultural and allied sectors are facing new challenges like the reducing availability of quality water, nutrient deficiency in soils, climate change, farm energy availability, loss of bio-diversity, emergence of new pests and diseases, rural-urban migration, besides globalisation of agri-food markets and trade regulations and these need to be addressed during the years to come. The farm sector stress and non-realisation of the desired/accelerated growth rate are attributable to several factors, some of which are as follows:

- Stagnant yields in last 20 years except in cotton – importing edible oils and yet to reach pulse – self-sufficiency.
- Depletion of critical resources – water & soil.
- Climate change – rising temperature and frequency of extreme weather events.
- Infrastructure constraints – power, roads, storage (cold & dry) and other agri-logistics.
- Low risk bearing capacity of farmers – poor farm returns.
- Low profitability in many crops / due to high vulnerability geographies – high cost of production; and less than remunerative returns.

High market risks – information asymmetry and market inefficiency.

- Inadequate financial services – credit and insurance.
- Increasing labour shortage and cost.
- Lack of data capture from field making price & demand forecasts (useful in decision making) difficult.
- Poor use of technology and modern science – as reflected in wide yield gaps between FLDs (frontline demonstrations) and farmers' fields.

In India, subsidies and increased awareness about fertilizers have led to a significant increase in fertilizer consumption. Importantly, while fertilizer consumption has continued to rise substantially, the elasticity of output with respect to fertilizer use has dropped sharply. During the period of 1970-71 to 2010-11, while foodgrain production grew by about 2.3 times, the increase in fertilizer (NPK) consumption was about 13 times (GoI, 2014). In the year 2013-14 food production was 264.8 million ton with use of 24.5 million tonnes of NPK fertilizers. The average crop response which was about 50 kg of foodgrain per kg of NPK fertilizer during the year 1970-71, fell to about 18.70 kg during 2010-11 and further down to about 10.8 during 2013-14. It also needs to be noted, that the increase in fertilizer use has come at significant cost. The fiscal burden of fertilizer subsidy which was just Rs 60 crore in 1976-77, shot up to over Rs 70,000 crore in 2012-13. It was as high as Rs. 72,437.58 crore in the year 2016-17. There are other important costs in the form of long-term soil degradation, degradation of water resources (in both quantity and quality), and general stagnation of yields due to application of sub-optimal nutrient ratios. Thus, disproportionate NPK fertilizer application, multi-nutrient deficiencies, and lack of organic manure application has led to reduction in the carbon content of the soil and contributed to stagnating agricultural productivity. This is turn has been getting reflected in high cost of production over the years.

References

Anonymous (2016). Report on Doubling Farmer's Income by 2022 Farm Crisis and Farmers' Distress 30 April, 2016, Indian Council of Food and Agriculture India International Centre, New Delhi.

Byerlee, D.; Diao, X.; Jackson, C. (2009). Agriculture, rural development, and poor growth: Country experiences in the post- reform era (Agriculture and Rural Development Working Paper 21). Washington, D.C.: World Bank.

Chadha, G.K.; Ramasundaram, P.; Sendhil, R. (2013). Current Science 105: 908-913.

Chand, R. (2016). Why doubling farm income by 2022 is possible. The ideas page, The Indian Express, April 15.

Chand, R. (2017). Doubling Farmers' Income: Rationale, Strategy, Prospects and Action Plan, NITI Policy Paper 01/2017 National Institution for Transforming India, Government of India, New Delhi.

Chand, R.; Saxena, R.; Rana, S. (2015): Estimates and Analysis of Farm Income in India, 1983-84 to2011-12, Economic and Political Weekly, Vol 32, May 30.

Chandrasekhar, S.; Mehrotra, N. (2016). Doubling Farmers' Incomes by 2022: What Would It Take? Economic & Political Weekly 51 pp.10-13.

Damodaran, H. (2017). Distress in farms: GDP data confirms deflation is here. The Indian Express, September 02, 2017Datt and Ravallion, 1998.

Department of Agriculture Cooperation & Farmers Welfare (DAC&FW) (2018). Ministry of Agriculture & Farmers Welfare, Government of India. www.agricoop.nic.in/doubling-farmers.

DFI (2018). Volume XII Science for Doubling Farmers' Income (DFI). Document prepared by the Committee for Doubling Farmers' Income, Department of Agriculture, Cooperation and Farmers' Welfare, Ministry of Agriculture & Farmers' Welfare.

DFI Committee (2017). Report of the Committee on Doubling Farmers Income, Vol II. Status of Farmers' Income: Strategies for Accelerated Growth. Department of Agriculture, Cooperation and Farmers' Welfare, Ministry of Agriculture & Farmers' Welfare.

Dholakia; Ravindra, H.; Manish, B.; Pandya; Payal M. Pateriya (2014). Urban – Rural Income Differential in Major States: Contribution of Structural Factors, W.P. No. 2014-02-07, Indian Institute of Management, Ahmedabad.

Doubling Farmers' Incomes: Strategies and Recommendations for Tamil Nadu NABARD, and IFMR 2017.

Fan, S.; Hazell, P.; Thorat, S. (1999). Linkages between government spending, growth, and poverty in rural India (Research Report No. 100). Washington, D.C.: International Food Policy Research Institute. Retrieved from http://www.ifpri.org/sites/default/files/publications/rr110.pdf

Gulati, A.; Saini, S. (2016): From plate to plough: Raising farmers' income by 2022, The Indian Express,April 12. http://goo.gl/2TKD61

http://www.agricoop.gov.in/sites/default/files/12%20July%202016%20Sat%20kjss%20NM.pdf

http://www.cicr.org.in/pdf/press/income-20-7-2018.pdf

https://blog.mygov.in/sweet-revolution-mithi-kranti-steps-towards-doubling-farmers-income/

https://economictimes.indiatimes.com/news/economy/agriculture/govt-to-double-farmers-income-by-2022-by-focusing-on-7-sources-of-income-says-economic-survey/articleshow/70074052.cms

https://economictimes.indiatimes.com/news/economy/agriculture/how-will-you-double-farmers-income-european-union-to-pm-modi/articleshow/69837021.cms?utm_source=contentofinterest&utm_medium=text&utm_campaign=cppst

https://niti.gov.in/writereaddata/files/document_publication/DOUBLING%20FARMERS%20INCOME.pdf

https://www.hindustantimes.com/india-news/centre-plans-wider-method-to-measure-income-of-farmers/story-xSQOdAEDCbGtsuW8lRefFL.html

Irz, X.; Lin, L,; Thirtle, C.; Wiggins, S. (2001). Development Policy Review 19 (4): 449-466.

Mellor, J. (1999). Faster, more equitable growth – The relation between growth in agriculture and poverty reduction agricultural policy development project (Research Report No. 4). Washington, D.C.: United States Agency for International Development.

Minten, B. Barrett, C. (2008). World Development 36 (5): 797-822.

Muyanga, M.; Jayne, T.; Burke, W.J. (2010). Pathways into and out of poverty: A study of rural household wealth dynamics in Kenya (Working Paper). Nairobi, Kenya: Tegemeo Institute of Agricultural Policy and Development.

NITI Aayog (2015). Raising agricultural productivity and making farming remunerative for farmers. Occasional Paper NITI Aayog, Government of India, pp. 1 - 46.

NSSO (National Sample Survey Office), 2104. Report No NSS KI (70/33).

Satyasai, K.J.S.; Sandhya, B. (2016). Doubling Farmers Income: way forward, Rural Pulse. Issue - XIV, Occasional Paper, National Bank for Agriculture and Rural Development (NABARD), Mumbai.

Saxena, R.; Singh, N.P.; Balaji, S.J.; Usha, R.A.; Deepika, J. (2017). Strategy for Doubling Income of Farmers in India. Policy Paper 31. ICAR-National Institute of Agricultural Economics and Policy Research (NIAP), New Delhi.

Sendhil, R.; Balaji, S.J.; Ramasundaram, P.; Kumar, A.; Singh, S.; Chatrath, R.; Singh. G.P. (2018) Doubling Farmers Income by 2022 Trends, Challenges, Pathway and Strategies, ICAR- Indian Institute of Wheat and Barley Research Karnal, Haryana.

Sendhil, R.; Kar, A.; Mathur, V.C.; Jha, G.K. (2013). Agricultural Economics Research Review 26(1): 41-54.

Sendhil, R.; Kar, A.; Mathur, V.C.; Jha, G.K. (2014a). Journal of Indian Society of Agricultural Statistics 68(3): 365-375.

Sendhil, R.; Ramasundaram, P.; Anbukkani, P.; Singh, R.; Sharma, I. (2015). Trends and Determinants of Research Driven Total Factor Productivity in Indian Wheat, International Conference of Agricultural Economists held at the University of Milan, Italy.

Sendhil, R.; Singh, R.; Ramasundaram, P.; Kumar, A.; Singh, S.; Sharma, I. (2014b). Journal of Wheat Research 6(2): 138-149.

Sharma, I.; Chatrath, R.; Sendhil, R. (2013). Indian Farming 63(8): 3-6.

Timmer, P. (1995). Food Policy 20(5): 455– 472.

Waghmare, A. (2016): Why it is hard to double farmers' income by 2022, March 30. http://goo.gl/mqZ27q

7

Technological Innovations and Practices

Technological change has been the major driving force for increasing agricultural productivity and promoting agriculture development across the world. In the past, the choice of technologies and their adoption was to increase production, productivity and farm incomes. Over many decades, policies for agriculture, trade, research and development, education, training and advice have been strong influences on the choice of technology, the level of agricultural production and farm practices. Agricultural production today faces several challenges: an increasing world population has to be fed, but at the same time the environment and animal welfare need to be protected. Globalisation has led to increasing international competition, while the agricultural sector is also under heavy economic pressure as a result of regulation. These conditions can perhaps only be met by innovation, be it on the technological or on the systems-level (OECD, 2001). Agriculture is becoming more integrated in the agro-food chain and the global market, while environmental, food safety and quality, and animal welfare regulations are also increasingly impacting on the sector. It is faced with new challenges to meet growing demands for food, to be internationally competitive and to produce agricultural products of high quality. At the same time, it must meet sustainability goals in the context of on-going agricultural policy reform, further trade liberalization and the implementation of multilateral environmental agreements as agreed to by the government.

Today, farmers, advisors and policy makers are faced with complex choices. They are faced with a wide range of technologies that are either available or under developing stages; they must deal with the uncertainties of both the effects these new technologies will have throughout the agri-food chain and the impact that a whole range of policies will have on the sustainability of farming systems. In addition, there is increasing pressure on agricultural research and advisory budgets that must be accommodated. The focus of the workshop was the adoption of technologies that have the potential to contribute to sustainable farming systems. Technology adoption, however, is a broad concept. It is affected by the development, dissemination and application at the farm level of existing and new biological, chemical and mechanical techniques, all of which are encompassed in farm capital and other inputs; it is also affected by education, training, advice and information which form the basis of farmers' knowledge. It also includes technologies and practices in the whole agri-food sector that have an impact at the farm level. Finally, it should be borne in mind that most of these new technologies originate outside the farm sector. The concept of a sustainable farming system refers to the capacity of agriculture over time to contribute to overall welfare by providing sufficient food and other goods and services in ways that are economically efficient and profitable, socially responsible, while also improving environmental quality. It is a concept that can have different implications in terms of appropriate technologies whether it is viewed at the farm level, at the agri-food sector level, or in the context of the overall domestic or global economy.

7.1 New Agriculture Technology in Modern Farming

Innovation is more important in modern agriculture than ever before. The industry as a whole is facing huge challenges, from rising costs of supplies, a shortage of labor, and changes in consumer preferences for transparency and sustainability. There is increasing recognition from agriculture corporations that solutions are needed for these challenges. In the last 10 years, agriculture technology has seen a huge growth in investment, with $6.7 billion invested in the last 5 years and $1.9 billion in the last year alone. Major technology innovations in the space have focused around areas such as indoor vertical farming, automation and robotics, livestock technology, modern greenhouse practices, precision agriculture and artificial intelligence, and blockchain (Linly Ku).

7.1.1 Indoor Vertical Farming

Indoor vertical farming can increase crop yields, overcome limited land area, and even reduce farming's impact on the environment by cutting down distance traveled in the supply chain. Indoor vertical farming can be defined as the practice of growing produce stacked one above another in a closed and controlled environment. By using growing shelves mounted vertically, it significantly reduces the amount of land space needed to grow plants compared to traditional farming methods. This type of growing is often associated with city and urban farming because of its ability to thrive in limited space. Vertical farms are unique in that some setups don't require soil for plants to grow. Most are either hydroponic, where vegetables are grown in a nutrient-dense bowl of water, or aeroponic, where the plant roots are systematically sprayed with water and nutrients. In lieu of natural sunlight, artificial grow lights are used.

From sustainable urban growth to maximizing crop yield with reduced labor costs, the advantages of indoor vertical farming are apparent. Vertical farming can control variables such as light, humidity, and water to precisely measure year-round, increasing food production with reliable harvests. The reduced water and energy usage optimizes energy conservation -- vertical farms use up to 70% less water than traditional farms. Labor is also greatly reduced by using robots to handle harvesting, planting, and logistics, solving the challenge farms face from the current labor shortage in the agriculture industry.

7.1.2 Farm Automation

Farm automation, often associated with "smart farming", is technology that makes farms more efficient and automates the crop or livestock production cycle. An increasing number of companies are working on robotics innovation to develop drones, autonomous tractors, robotic harvesters, automatic watering, and seeding robots. Although these technologies are fairly new, the industry has seen an increasing number of traditional agriculture companies adopt farm automation into their processes.

New advancements in technologies ranging from robotics and drones to computer vision software have completely transformed modern agriculture. The primary goal of farm automation technology is to cover easier, mundane tasks. Some major technologies that are most commonly being utilized by farms include: harvest automation, autonomous tractors, seeding and weeding, and drones. Farm automation technology addresses major issues like a rising global population, farm labor shortages, and changing consumer preferences. The benefits of automating traditional farming processes are monumental by tackling issues from consumer preferences, labor shortages, and the environmental footprint of farming.

7.1.3 Livestock Farming Technology

The traditional livestock industry is a sector that is widely overlooked and under-serviced, although it is arguably the most vital. Livestock provides much needed renewable natural resources that we rely on every day. Livestock management has traditionally been known as running the business of poultry farms, dairy farms, cattle ranches, or other livestock-related agribusinesses. Livestock managers must keep accurate financial records, supervise workers, and ensure proper care and feeding of animals. However, recent trends have proven that technology is revolutionizing the world of livestock management. New developments in the past 8-10 years have made huge improvements to the industry that make tracking and managing livestock much easier and data-driven. This technology can come in the form of nutritional technologies, genetics, digital technology, and more.

Livestock technology can enhance or improve the productivity capacity, welfare, or management of animals and livestock. The concept of the 'connected cow' is a result of more and more dairy herds being fitted with sensors to monitor health and increase productivity. Putting individual wearable sensors on cattle can keep track of daily activity and health-related issues while providing data-driven insights for the entire herd. All this data generated is also being turned into meaningful, actionable insights where producers can look quickly and easily to make quick management decisions.

Animal genomics can be defined as the study of looking at the entire gene landscape of a living animal and how they interact with each other to influence the animal's growth and development. Genomics help livestock producers understand the genetic risk of their herds and determine the future profitability of their livestock. By being strategic with animal selection and breeding decisions, cattle genomics allows producers to optimize profitability and yields of livestock herds. Sensor and data technologies have huge benefits for the current livestock industry. It can improve the productivity and welfare of livestock by detecting sick animals and intelligently recognizing room for improvement. Computer vision allows us to have all sorts of unbiased data that will get summarized into meaningful, actionable insights. Data-driven decision making leads to better, more efficient, and timely decisions that will advance the productivity of livestock herds.

7.1.4 Modern Greenhouses

In recent decades, the Greenhouse industry has been transforming from small scale facilities used primarily for research and aesthetic purposes (i.e., botanic gardens) to significantly more large-scale facilities that compete directly with land-based conventional food production. Combined, the entire global greenhouse market currently produces nearly US $350 billion in vegetables annually, of which U.S. production comprises less than one percent. Now-a-days, in large part due to the tremendous recent improvements in growing technology, the industry is witnessing a blossoming like no time before. Greenhouses today are increasingly emerging that are large-scale, capital-infused, and urban-centered.

As the market has grown dramatically, it has also experienced clear trends in recent years. Modern greenhouses are becoming increasingly tech-heavy, using LED lights and automated control systems to perfectly tailor the growing environment. Successful greenhouse companies are scaling significantly and located their growing facilities near urban hubs to capitalize on the ever-increasing demand for local food, no matter the season. To accomplish these fects, the greenhouse industry is also becoming increasingly capital-infused, using venture funding and other sources to build out the infrastructure necessary to compete in the current market.

7.1.5 Precision Agriculture

Agriculture is undergoing an evolution - technology is becoming an indispensable part of every commercial farm. New precision agriculture companies are developing technologies that allow farmers to maximize yields by controlling every variable of crop farming such as moisture levels, pest stress, soil conditions, and micro-climates. By providing more accurate techniques for planting and growing crops, precision agriculture enables farmers to increase efficiency and manage costs. Precision agriculture companies have found a huge opportunity to grow. A recent report by Grand View Research, Inc. predicts the precision agriculture market to reach $43.4 billion by 2025. The emerging new generation of farmers are attracted to faster, more flexible startups that systematically maximize crop yields.

7.1.6 Blockchain

Block chain's capability of tracking ownership records and tamper-resistance can be used to solve urgent issues such as food fraud, safety recalls, supply chain inefficiency and food traceability in the current food system. Block chain's unique decentralized structure ensures verified products and practices to create a market for premium products with transparency. Food traceability has been at the center of recent food safety discussions, particularly with new advancements in blockchain applications. Due to the nature of perishable food, the food industry at whole is extremely vulnerable to making mistakes that would ultimately affect human lives. When foodborne diseases threaten public health, the first step to root-cause analysis is to track down the source of contamination and there is no tolerance for uncertainty. Blockchain can be used to solve urgent issues such as food fraud, safety recalls, supply chain inefficiency and food traceability in the current food system. Consequently, traceability is critical for the food supply chain. The current communication framework within the food ecosystem makes traceability a time-consuming task since some involved parties are still tracking information on paper. The structure of blockchain ensures that each player along the food value chain would generate and securely share data points to create an accountable and traceable system. Vast data points with labels that clarify ownership can be recorded promptly without any alteration. As a result, the record of a food item's journey, from farm to table, is available to monitor in real-time.

The use cases of blockchain in food go beyond ensuring food safety. It also adds value to the current market by establishing a ledger in the network and balancing market pricing. The traditional price mechanism for buying and selling relies on judgments of the involved players, rather than the information provided by the entire value chain. Giving access to data would create a holistic picture of the supply and demand. The blockchain application for trades might revolutionize traditional commodity trading and hedging as well. Blockchain enables verified transactions to be securely shared with every player in the food supply chain, creating a marketplace with immense transparency.

7.1.7 Artificial Intelligence

The rise of digital agriculture and its related technologies has opened a wealth of new data opportunities. Remote sensors, satellites, and UAVs can gather information 24 hours per day over an entire field. These can monitor plant health, soil condition, temperature, humidity, etc. The amount of data these sensors can generate is overwhelming, and the significance of the numbers is hidden in the avalanche of that data. The idea is to allow farmers to gain a better understanding of the situation on the ground through advanced technology (such as remote sensing) that can tell them more about their situation than they can see with the naked eye. And not just more accurately but also more quickly than seeing it walking or driving through the fields. Remote sensors enable algorithms to interpret a field's environment as statistical data that can be understood and useful to farmers for decision-making. Algorithms process the data, adapting and learning based on the data received. The more inputs and statistical information collected, the better the algorithm will be at predicting a range of outcomes. And the aim is that farmers can use this artificial intelligence to achieve their goal of a better harvest through making better decisions in the field.

7.2 Effective Microorganisms (EMs) Technology

Effective Microorganisms (Ems) is a consortium culture of different effective microbes commonly occurring in nature. Most important among them are: N_2-fixers, P-solubilizers, photosynthetic microorganisms, lactic acid bacteria, yeasts, plant growth promoting Rhizobacteria and various fungi and Actinomycetes (NCOF, 2017). In this consortium, each microorganism has its own beneficial role in nutrient cycling, plant protection and soil health and fertility enrichment. EM are mixed cultures of beneficial naturally-occurring organisms that can be applied as inoculants to increase the microbial diversity of soil ecosystem. They consist mainly of the photosynthesizing bacteria, lactic acid bacteria, yeasts, actinomycetes and fermenting fungi. These microorganisms are physiologically compatible with one another and can coexist in liquid culture. Several other authors Kengo and Hui-lian, (2000); Zuraini *et al.,* (2010); Zakaria *et al.,* (2010); David *et al.,* (2010); Allahverdiyev *et al.,* (2011) Rahmann and Aksoy, (2014); Bargaz *et al.,* (2018); Bzdyk *et al.,* (2018) and Naik *et al.,* (2020) reported the importance to EM in various ways.

7.2.1 EM Formulations

7.2.1.1 EM-1 Formulation

This formulation is used for seed treatment, soil enrichment and for spray in field after the emergence of seedlings.

- Dissolve 5 kg jaggary (chemical free) in about 100 lit of water.
- Add 5 lit of EM.
- Mix thoroughly and pour into a plastic carboy (a large bottle for holding corrosive liquid). Seal the carboy and allow to ferment for 7 days.
- Dilute this solution in a ratio of 1:1000 and spray over soil or crop residue. For seed treatment soak the seeds in this diluted solution.

7.2.1.2 EM-5 for control of insects and pests

- Dissolve 100gm of jaggary in 600 ml of water.
- Add 100 ml each of natural vinegar, wine or brandy and EM.
- Mix thoroughly and transfer the contents in a plastic bottle or carboy and seal the container.
- To increase the potency few cloves of garlic and chilly paste can also be added to this suspension before sealing the container.
- Allow the contents to ferment for 5-10 days under shade.
- Release the gas daily.
- Within 10 days the EM solution will be ready for use. This can be stored up to 3 months at normal room temperature in a cool and dry place.
- Dilute the contents in a ratio of 1: 1000 and apply as foliar spray with the help of a sprayer.

7.2.2 How to use EM: Application of EM in agriculture involves four steps as follows:

a. Procurement of primary EM- available in market.
b. Preparation of secondary EM – to be carried out by the farmer.
c. Appropriate dilution of the secondary EM solution.
d. Application to plants, soil and organic matter as spray.

Preparation of secondary EM solution depending upon the requirement and its end use, various EM formulations have been developed. Even among one formulation depending upon the place and climatic conditions some variations have been incorporated and recommended by promoting institutes and agencies. Some of the widely used and popular formulations are described below. Water used in all formulations should be either rainwater or fresh tube-well water. Tap water is not to be used.

7.2.3 Benefits of EM

- **EM as a compost:** EM can be applied to the compost heap to reduce troublesome odours and flies as well as improving the compost process and quality. Preferably spray on with a hand sprayer to prevent over wetting the compost heap and apply at each addition of fresh material if possible.
- **EM cultures and organics:** EM cultures have been used effectively to inoculate both farm wastes as well as urban wastes to reduce odours and hasten the treatment process. EM has also been used with great success as an inoculant for composting a wide variety of organic wastes. An EM culture known as EM Bokashi can be used for composting food organics and other compostable materials. EM Bokashi is a fermented compost starter made from sawdust and wheat bran. When the correct conditions are provided EM sets in motion a fermentation process to transform food and other organic materials into compost.
- **EM effects on soils and crops:** EM has been used on many different soils and crops over a wide range of conditions. EM enhances soil fertility and promotes growth, flowering, fruit development and ripening in crops. It can increase crop yields and improve crop quality as well as accelerating the breakdown of organic matter from crop residues. EM cultures can accelerate the decomposition of organic wastes, increase the availability of mineral nutrients and useful organic compounds to plants, and enhance the activities of beneficial micro-organisms, e.g., mycorrhizae and nitrogen fixing bacteria.
- **EM for weeds pests and diseases:** EM is not a pesticide and contains no inorganic chemicals. EM is a microbial inoculant that works as a bio-control measure in suppressing and/or controlling pests through the introduction of beneficial microorganisms to soils and plants. Pests and pathogens are suppressed or controlled through natural processes by enhancing the competitive and antagonistic activities of the microorganisms in the EM inoculants. Thus, EM helps to increase beneficial soil micro-organisms and suppression of harmful ones.

- **EM as promoter for VAM:** Interdependent biological activity of different EM organisms creates a congenial environment for growth and spread of soil's flora and fauna. They also promote the growth and colonization of VAM, which further help in plant growth promotion.
- **EM the Natural Product:** EM is a combined culture of aerobic microorganisms (requiring oxygen to survive) and anaerobic (requires no oxygen to survive) that co-exist together to the mutual advantage of both (symbiosis).

7.3 Fermented Plant Extract (FPE)

In this formulation fresh green weeds are fermented with EM to obtain a fermented plant extract.

- Cover the drum and tie with a rope.
- The drum should be filled up to the top, leaving very little space for air.
- Fermentation and gas formation process will start slowly.
- Mix the contents at repeated intervals.
- Finished FPE having a pH of 3.5 with pleasing smell will be ready in 5-10 days' time.
- Filter the solution through a cloth and collect the filtrate.
- For spraying on soil dilute the FPE in a ratio of 1: 1000 with fresh water.
- For spraying on crops dilute FPE in a ratio of 1: 500.
- Spraying should be done after germination of seeds in early morning hours once or twice a week.

7.4 EM – Bokashi

Bokashi is a type of compost prepared by fermentation of waste organic matter with the help of EM. Bokashi is mainly used for improving the fertility status of soil and for enhancing the degradation of crop residue.

- Collect sufficient quantity of different organic matter (such as rice bran, fish meal, animal waste etc,) equivalent to 150 lit drum volume.
- Mix 150gm of jaggary and 50 ml of EM in 15 lit of water.
- Mix this solution with organic waste thoroughly in such a way that entire contents get uniformly moistened.
- Grind 2.3kg of fresh green weeds to a coarse paste. Dilute with 14 lit of water.

- Dissolve 42gm of jaggary in some water and mix with weed suspension.
- Add 420ml of EM.
- Transfer the contents to a plastic drum and with the help of a thick plastic sheet
- Transfer the contents in a plastic bag and seal the bag.
- To ensure the anaerobic conditions put this bag into another polythene bag and seal
- Allow the contents to ferment for 3-4 days in a cool shade place
- Bokashi will be ready after 4 days.
- This can be used immediately.
- In plastic air tight bags Bokashi can be stored up to 6 months.

How to use Bokashi

Bokashi can be used directly as compost in poor fertility soils. It can also be used along with the crop residues. For 0.1 ha mix 100-150 kg Bokashi with sufficient quantity of finely chopped crop residue. Spread this mixture over 0.1 ha area and mixed with soil a day before sowing. Spraying of 5-10 lit of 1:500 diluted simple EM-solution over this mixture can further boost the degradation process. By using Bokashi + crop residue + EM-solution the requirement of compost can be dispensed with. This can save lot of labour, time and space required for compost process.

7.5 Indian Agritech Startups (Bhatia, 2018)

AgriTech in India has become a booming field with numerous startups working with technologies such as data analytics, machine learning and satellite imaging, among others, enabling farmers to maximise their output. According to a NASSCOM report, the Indian Government specifically supports AgriTech startups through its Startup India program. In 2016, more than 350 AgriTech startups raised $300 million in investment globally, out of which Indian investment accounted for 10 percent. AgriTech is clearly leading India's next green revolution. Himanshu Goyal, India sales and alliances leader at The Weather Company said, "IBM sees a INR 5000.00 crore opportunity from AgriTech in India over the next five years." A clutch of tech incubators in India are also supporting innovation in agriculture, especially in the rural areas. Case in point is Indigram Labs, a technology-driven business incubator that aims to incubate 100 AgriTech entrepreneurs over the next five years.

7.5.1 SatSure: Founded in early 2016, SatSure has been at the forefront of bringing the best practices of satellite image processing, big data capabilities, and IT to agriculture. It also strives to create a positive impact on the lives of farmers by helping improve crop insurance, innovate on agri lending services, and improve market linkages by creating intervention and decision intelligence frameworks for agri value chain stakeholders. The startup has mobile app platforms for delivering information on supply statistics of crops and crop stressing in their region. This helps with the decisions on what to sow, when to irrigate or add fertilizers, or prepare for harvest. Currently, the startup's solutions are being used by the Andhra Pradesh Government. Large banks and insurance companies in India are also leveraging SatSure's solutions. NITI Aayog supported the startup through initiatives like the Grand Agriculture Challenge which further validates our belief the data mining and analysis is of high value to the agriculture sector.

7.5.2 Fasal: Founded by Shailendra Tiwari and Ananda Verma, Fasal's microclimate forecasts are tailored to each farm location and are performed at a point scale, not at a kilometer-wide spatial scale. Fasal founder explained that as the startup collects more data, the AI-based microclimate forecasting algorithm incorporates real in-field information and relates it to publicly available weather forecasts, so that farmers can benefit from real-time, actionable information relevant to day-to-day operations at the farm.

7.5.3 Aibono: As India's first smart farming collective, this startup turns around the fortunes of small farmers with the internet, AI, shared services. Pegged as Agri 4.0, the collective provides precision agriculture technologies backed by real-time synchronisation of supply and demand. Some of its solutions include real-time precision agriculture as a service to farmers. The company has about 60 employees and will be 100-strong by the end of the year 2018.

7.5.4 Gobasco: This Gurgaon and Lucknow-based AgriTech startup leverages real-time data analytics on data-streams coming from multiple sources across the country and is backed by AI-optimised automated pipelines to improve and increase the efficiency of the agricultural supply chain. The startup's data-driven online agri-marketplace gives the best prices for both the producers and buyers at their fingertips. Some of its solutions include transaction discovery, procurement optimisation and optimising transportation with real-time data. Last year, the startup raised an undisclosed amount in seed funding from Matrix Partners.

7.5.5 Cropin: Bengaluru-based Cropin provides a full suite of farm management, monitoring and analytics solutions. Last year, the company invested $5 million toward AI by developing a new product called SmartRisk.

SmartRisk is essentially a digital platform for microfinance, banking and non-banking institutions to identify and minimise the risk in lending and insurance business. As per a company statement, SmartRisk application will be positioned along with the flagship SmartFarm application, which currently assists over five million farmers toward farm management, crop cycle monitoring, harvest and brings in produce traceability from farm to fork. SmartFarm also provides holistic farm management, while SmartSales benefits agro-input companies and Warehouse lays down the norms of food traceability to the last mile.

7.5.6 Intello Labs: Positioned as India's most awarded AgriTech startup, this Bengaluru-based startup has developed computer vision based solutions that use images as key data for deriving insights and actionable recommendations. The two key agri products from Intello Labs are used for crop inspection and agricultural product grading. Both the products read images and give quality parameters based on the input data. The startup leverages emergent technologies — deep learning, AI and IoT to help farmers scale their business effectively.

7.6 Some Technological Innovations

Some of the technological innovation are given here those are revolutionizing Indian Agriculture (Reddy, 2017)

7.6.1 Barrix Agro Sciences

The Bangalore-based startup oers eco-friendly crop protection methods after much research on products that support organic farming to increase crop produce and quality with minimal expenditure. **Products:** Barrix Catch Fruit and Fly Lure + trap: Toxic pesticides contaminate water, soil and leave behind harmful residue, besides being expensive. Barrix's pheromone-based pest control traps have artificially synthesized smelling agents that attracts and traps pests. Instead of eating the crops, the pests are attracted to the pheromones in the trap. Fly pest sticky sheet: Barrix uses bright yellow and blue coloured recyclable sheets of wavelengths between 500 nm to 600 nm, proven to effectively attract and trap at least 19 high-risk pests from a long distance

7.6.2. Anulek Agrotech

Set up by Mumbai-based entrepreneurs, Anulekh focuses on increasing soil fertility to achieve higher agricultural productivity and crop yield with lower resource use. **Product:** BIOSAT: BIOSAT (Biochar based Organic Soil Amendment Technology), a soil additive, is made of biochar mixed with different organic nutrients. The product preserves soil fertility, traps carbon emissions, maintains the top soil strength and increases crop production, thus reducing dependency on chemical fertilizers.

7.6.3 Mitra

A Nashik-based startup, MITRA (Machines, Information, Technology, Resources for Agriculture) aims to improve mechanization at horticulture farms with the use of R&D and high quality farm equipment. **Products:** Air blast sprayers: Developed for fruits and vegetables in general, and grapes and pomegranates in particular, the sprayers, used to add hormones that help the growth of crops, reduce the expenditure on manual labour and are less time-consuming.

7.6.4 CropIn Technology Solutions

A farming technology solutions startup founded by a Bangalore software engineer, it provides agri businesses, the technology and expertise to create a smarter and safer food supply for consumers around the world. **Product:** CropIn offers information on a cloud-based platform, integrated with a mobile app for Android. Called Smart Farms, it allows large food companies to track the growth of crops on farms around the country with details about what the crop is and the conditions it is grown in to help companies remotely monitor farms, interact with farmers and make every crop transparents and traceable. It also aids farmers in adopting global agricultural practices and improves productivity by offering productivity insights and harvest forecasts.

7.6.5 Eruvaka Technologies

An organisation based in Vijayawada, Andhra Pradesh, its mission is to accelerate the use of technology in aquaculture, an area where farmers face problems due to unavailability of adequate technology to measure and control water health. **Product:** Eruvaka Technologies, to help farmers monitor aquaculture ponds, develops solar-powered floating buoys that measure different water parameters, such as oxygen levels, temperature and pH range, crucial for the growth and survival of fish and shrimp. The collected information is uploaded on the cloud and transmitted to individual customers through an Android app, SMS, voice call or the internet. Farmers can also remotely control automated equipment such as aerators and feeders.

7.6.6 Skymet

Skymet is India's largest weather monitoring and agri-risk Solutions Company. According to their website, they are the experts in measuring, predicting, and limiting climate risk to agriculture, thus reducing losses incurred due to bad weather conditions. **Product:** Launched to aid farmers, Skymet's weather website offers services such as weather forecast, crop insurance and agri-risk management. Prediction of weather conditions can help prepare a farmers for a drought or heavy unseasonal rainfall and help them take appropriate preventive measures, they say and claim to accurately measure and predict yield at the village level for any crop.

7.6.7 Ekgaon

Gujarat-based venture started in 2001, Ekgaon Technologies is an IT based network integrator that provides a technology platform and offers a range of services to farmers in rural areas including financial, agricultural inputs and government assistance. **Product: (a)** Financial: A mobile phone enabled financial services delivery platform, it provides information on microfinance institutions and banks for delivery of door-step services such as credit, savings, remittance, insurance, investment and mortgage. (b) **Agricultural:** Offered in Hindi, Gujarati and Tamil languages, the system uses mobile, voice recognition, interactive voice response system (IVRS) and web technologies to provide information on weather, commodity market prices, soil nutrient management and crop management. (c) **Citizen:** The web and mobile applications help citizens monitor the delivery of government programmes and services entitled to them.

7.6.8. Digital Green

Digital Green is a not-for-profit international development organisation that focuses on training farmers to make and show short videos where they record their problems, share solutions and highlight success stories as community engagement to improve lives of rural communities across South Asia and Sub-Saharan Africa. **Product:** It uses technology-enabled behaviour change communication that is cost-effective, scalable and brings together researchers, development practitioners, and rural communities to produce and share locally relevant information through videe. Two social online games Wonder Village and Farmer Book: In the games, players simulate a village economy and relate with actual farmers that Digital Green works with, on the field. The players are placed in a resource-constrained setting in which they have to complete quests such as set up paddy and maize farms and supply raw materials to the farmers' markets.

7.6.9. Frontal Rain Technologies

The Bangalore-based agri-tech startup seeks to deliver affordable advanced technology solutions for emerging companies and take technology to remote corners of the country. **Product:** The Company's offering Rain+, according to their website, is a comprehensive suite of products on the cloud for food and agribusinesses. Rain+ can help companies at every stage of the value chain starting from growing, processing, logistics, wholesale trade, retail trade and exports. This technology, accessible through desktop, tablet and mobile devices, is used by companies dealing with commodities like spices, herbs, basmati rice, seeds, animal feed, sea food, dairy and edible oil.

7.6.10. Agro Star

A Pune-based 'direct to farmer' m-commerce platform, Agrostar strives to provide quality agro inputs at the farmers' doorstep. **Product:** AgroStar enables farmers to procure a range of agricultural goods such as seeds, crop nutrition, crop protection and agri-hardware products by simply giving a missed call on the company's 1800 number or through their mobile app to eliminate unavailability of products, substandard products, duplication and adulteration.

7.7 Farmers' Innovation (Source: TNAU Agritech Portal)

7.7.1. A Young Innovator: Solar Seeder

Solar Seeder is innovated by M. Subash Chandra Bose Pudukkottai, Tamil Nadu. He participated in the National Level Science Competition Ignite – 2015 held at IIM Ahmedabad. He received 'Dr APJ Abdul kalam Ignite Award-2015' from Honorable President Shri Pranab Mukherji for his innovation "Solar Seeder".

Benefits of using Solar Seeder

- Uniform gap is maintained and wastage of seeds will be avoided.
- Since digging is not deep, soil moisture can be maintained.
- Total weight of the device approximately 200kgs, so it will not do harm to the ploughed land, soil package and structure.
- Since it was operated in Solar Energy there are no chances for environmental pollution due to emissions.
- Saves labour used in sowing operation
- It can sow the seeds like black gram, groundnut, green gram and chick peas.

7.7.2 Tapioca Chips cutting Machine

The Tapioca (Cassava) was one of the major crops cultivated in Salem, Namakkal and other parts of Tamil Nadu. It is grown for industrial and edible purpose. During, 2011-12, the sale price of tuber reduced to lower rate which is uneconomical to farmers. In this back ground, the officials from state government (Agriculture) took an initiative to protect the interest of tapioca farmers. And the solution to the problem was *in situ* processing / value addition of tapioca, and this in turn has helped the tapioca farmers and the biological deterioration of starch has minimized to larger extent. This has stimulated Mr. Dhanraj to innovate a cassava chips cutting machine which had easen the task of value addition. This was welcomed by other farmers and the district collector has appreciated his noval idea and he was awarded by National Innovation Foundation –SRISTI SAMAN award for his innovation

Uses of Cassava Chips cutting Machine

- It can be transported from one place to another quickly.
- Needs minimum labour for processing.
- The chips thickness can be altered from 1 mm to 25 mm (at uniform size).
- Locally made with available materials.
- The chips can be used for cattle feed without any contamination (manually cut chips can be contaminated with - sand, iron, etc).
- There are very low chances of accident.
- The electricity driven motor can be replaced with oil engine (petrol or diesel engine) hence the capacity can be increased.
- It can be used for cutting / slicing vegetables at home or industry
- It is gender friendly
- The mangoes which are fallen off are collected sliced and preservatives (common salt) are added and dried and converted into edible purpose

7.7.3 Honey Wax Separator

In traditional methods of separation of wax from honey comb is a laborious process and time consuming. It eliminates smoke and creates polluted environment and the wax is deposited in the vessel. In order to overcome this problem a solar based wax separator developed by Master Jawahar Raja who is studying Xth std. this separator consists of steel box with reflected convex lens on both sides. The steel box is painted black in clolour inside and covered by thermocole sheath. The steel box has a small hole on the bottom in which the melted wax collected in the vessel which is fixed in the separate chair. Finally along with the wax remaining honey is collected. By adopting this technology to separate 10 kg of honey comb expected to get 7 kg of wax and 1 kg honey. An Individual can get additional income of Rs.2,400/- when compared to Rs.1,700/- in traditional method, hence we can save Rs.700.

7.7.4 Clod Breaker

Clod breaker was innovated by Shri G.R.Sakthivel. It is made up of wood in cylinder shape with 1 feet diameter and 4 feet length. Iron pokes 10mm length should be installed all over the wood log. The equipment should be made such that the wood log is supported and dragged by bullocks on the both sides and driven by a man seated. It is an animal drawn secondary tillage implement used for breaking the hard soil. The equipment has to be driven two times in an acre field for breaking the clods and the time consumed for this operation is five hours. It increase the water holding capacity of the soil and help for ensuring the germination of all the small seeded crops suitable for dry lands.

7.7.5 Fermented Castor Solution Trap

It was innovated by Shri G.R.Sakthivel, The mud pot with the castor solution needs to be buried near the trunk of the tree and attracted by the odour from the pot the insect comes towards it and fall into the pot and die. Collect the dead insects float found in the pot once in two days and filling the pots with solution whenever the quantity gets reduced. The solution can be kept for a period of three months. Fermented castor solution trap is effectively used for controlling the pests like white grub, stem weevil and Rhinoceros beetle. It also caused the no-entry / re-infestation of rats into the field. It is an eco-friendly cost effective techniques for controlling the white grub in sugarcane, cotton and groundnut crops. This method is found to reduce the plant protection expense to 20% since the cost of the entire process comes to Rs.250.00 only

References

Allahverdiyev, S.R.; Kırdar, E.; Gunduz, G.; Kadimaliyev, D.; Revin, V.; Filonenko, V.; Rasulova D.A.; Abbasova, Z.I.; Gani-Zade, S.I.; Zeynalova. (2011). E.M. Technology, 14(4): 103-106.

Bargaz, A.; Lyamlouli, K.; Chtouki, M.; Zeroual, Y.; Dhiba, D. (2018). Front Microbiol. 9: 1606.

Bhatia, R. (2018). Top 6 Indian Agritech Startups that are revolutionizing agriculture retrieved from https://analyticsindiamag.com/top-6-indian-agritech-startups-that-are-revolutionising-agriculture/

Bzdyk , R.M.; Jacek Olchowik , Marcin Studnicki , Tomasz Oszako, Katarzyna Sikora, Hanna Szmidla and Dorota Hilszcza ´nska (2018). The Impact of Effective Microorganisms (EM) and Organic and Mineral Fertilizers on the Growth and Mycorrhizal Colonization of Fagus sylvatica and Quercus robur Seedlings in a Bare-Root Nursery Experiment Forests 2018, 9, 597; doi:10.3390/f9100597 pp. 1-13.

David A. Swayne, Wanhong Yang, A. A. Voinov, A. Rizzoli, T. Filatova (Eds.) Modelling for Environment's Sake, Fifth Biennial Meeting, Ottawa, Canada retrieved from http://www.iemss.org/iemss2010/index.php?n=Main.Proceedings

http://agritech.tnau.ac.in/farm_innovations/farm_innovations.html

https://www.worldgovernmentsummit.org/api/publications/document?id=95df8ac4-e97c-6578-b2f8-ff0000a7ddb6

Kengo, Y. and Hui-lian, X. (2000). Journal of Crop production 3(1): 255-268.

Linly Ku retrieved from https://www.plugandplaytechcenter.com/resources/new-agriculture-technology-modern-farming/

Naik, K., Mishra, S., Srichandan, H. (2020). Sustain Environ Res 30, 10 (2020). https://doi.org/10.1186/s42834-020-00051

NCOF (2017). Organic Agriculture (Concept, Scenario, Principles and Practices) National Project on Organic farming Deptt of Agriculture and Cooperation, Govt. of India.

OECD (2001). https://www.oecd.org/greengrowth/sustainable-agriculture/2739771.pdf

Rahmann, G; Aksoy, U. (2014). Proceedings of the 4th ISOFAR Scientific Conference. 'Building Organic Bridges', at the Organic World Congress 13-15 Oct., Istanbul, Turkey (eprint ID 24318).

Reddy, S.S. (2017). 10 Technological Innovations that are revolutionizing Indian Agriculture retrieved from https://medium.com/@shyam052090/10-technological-innovations-that-are-revolutionizing-indian-agriculture-b7e73bce19a4

Zakaria, Z.; Gairola, S.; Mohd Shariff, N. (2010). Effective Microorganisms (EM) Technology for Water Quality Restoration and Potential for Sustainable Water Resources and Management International Environmental Modelling and Software Society (iEMSs).

Zuraini, Z.; Sanjay, G.; Noresah. M. (2010). Effective Microorganism (EM) technology for water quality restoration and potential for sustainable water resources and management. Proceedings of the International Congress on Environmental Modelling and Software Modelling for Environment's Sake, Fifth Biennial Meeting held between 5th- 8th July 2010, Ontario Canada.

8

Transformational Role of Digital Technologies Towards Agriculture

India is on the progressive path of development and its rural population is an integral part of this growth trajectory. As India gears up for an era of increased digitalisation, the issue of holistic and inclusive economic growth remains a pivotal concern. Hitachi, one of the leading Japanese conglomerate with a global footprint and a forerunner in digital innovation, has been an active contributor in transforming millions of Indian lives, its services reaching far beyond citizens within city limits. While India is one of the world's fastest-developing economy, equitable growth remains a critical imperative. The rural population today constitutes above 45% of the national income. In 2050, despite urbanization, over half of India's population will still be rural. Prime Minister, Sh. Narendra Modi has prioritized radical digitalisation to induce economic inclusiveness through a host of initiatives. 'Digital India', 'Make in India' and 'Skill India' provide for impetus and opportunity to rural citizens, to ensure they are equal participants in India's growth story. Digital transformation is the integration of digital technology into all areas of a business, fundamentally changing how you operate and deliver value to customers. It's also a cultural change that requires organizations to continually challenge the status quo, experiment, and get comfortable with failure. In many developed countries, farming has been modernized by a wave of technologies, adopted at farm level.

Digital technologies are finding increasing use in the agricultural value system, and High-tech farming is becoming the standard, thanks to use of sensors, logic controlled systems, data analytics, etc. In India, the increasing availability of energy and internet connectivity to the large rural landscape is further accelerating such changes. Digitised information systems allow remote access to knowledge, has given rise to group sharing and continue even now to revamp how societal exchanges happen interpersonally, commercially and in the extension services system. The digital spectrogram can be compared against a large database and spectral analysis can diagnose the contents of the sample. This analysis can happen in the cloud, and results communicated to the farmers almost instantaneously. There exist various similar technologies where a sample from a field, gets converted into digital information, is promptly analysed to provide accurate results, which then allows farmers to take decisions best suited to the land they farm on. The availability of satellite imagery, infrared imagery, and a myriad of remote inputs allow for more accurate weather forecasts, advance warnings on pest infestation and similar, and more. Instead of traditional homilies about temperature and rainfall events, the farmers now consult and share the advisories and forecasts of the same.

8.1 Digital Agriculture

Digital agriculture refers to tools that digitally collect, store, analyze, and share electronic data and/or information along the agricultural value chain. Sometimes known as "smart farming" or "e-agriculture," digital agriculture includes (but is not limited to) precision agriculture. Digital agriculture will facilitate the achievement of Sustainable Development Goals before 2030 and help the objectives of the National Food Security Act efficiently, effectively and in equitable manner so that everybody can access safe, nutritious and affordable food. (Bergvinson David, 2017). Startup India launched in 2016 aims to boost startups across sectors by providing handholding services, access to funding and incubation and is of great significances for the agriculture sector. The distinct vision of our Prime Minister assures regarding several initiatives taken to provide "Protective shield" to the farmers to increase production, Improve storage and connectivity with the consumers for better supply and profit (Ganguly and Patra, 2017).

Unlike precision agriculture, digital agriculture impacts the entire agri-food value chain before, during, and after on-farm production. Therefore, on-farm technologies, like yield mapping, GPS guidance systems, and variable-rate application, fall under the domain of precision agriculture and digital agriculture. On the other hand, digital technologies involved in e-commerce platforms, e-extension services, warehouse receipt systems, blockchain-enabled food traceability systems, tractor rental apps, etc. fall under the umbrella of digital agriculture but not precision agriculture. Geo-tagging of land, bio-tagging of livestock, bar-coding of planting material, and such, others methods are using digital technologies, for identifying and managing farm assets. Similarly, use of Aadhaar to uniquely identify an individual and manage the delivery of support, is another transforming intervention in farming. Digitalised, Direct Benefit Transfer (DBT) to farmers is a vital step, as will be the building of a comprehensive digital database of farmers. The roles played by digital technologies keep getting upgraded and there is need to continuously monitor and identify new developments, so as to prioritise applications for the purpose of enhancing farmers' income.

8.2 Digital Divide

Digitalization of the agrifood system involves the risk that the potential benefits will be unequally distributed between rural and urban areas, gender, and youth population. Urban areas often have better developed 'digital ecosystems' (resources, skills, networks) compared with rural areas. Combined with global trends of urbanization and middle and rich classes settling in cities, there is potential for digitalization to exacerbate existing rural urban disparities (UN DESA, 2018a) and populations to fall behind in the process of a digital transformation. FAO is committed to assist governments and partners bridging such multidisciplinary digital divides to ensure that everyone benefits from the emerging digital society (FAO, 2019).

8.3 Components for Modern Management of Agriculture

Digital technologies also allow for large amount of historic and real-time information to be recorded, shifted and correlated and this in turn optimises how information is used. Both individually and cumulatively, these technologies possess the power of ushering in a constructive disruption, a phrase widely popular by now. One of the pre-requisites for enhanced and stable farm income is sustainable and efficient management of agriculture yield and output. Management of diverse crop growth ambience, uncertainties of climate, soil and water regime will require pertinent and timely crop and soil information on temporal and spatial basis. Thus, a farmer needs to be informed well in advance of the probable upcoming problems and outbreaks. The possible components for modern management of agriculture are:

8.3.1 Remote Sensing

Remote sensing fundamentally made use of visible, near infrared and short-wave infrared sensors to form images of the earth's surface by detecting the solar radiation reflected from targets on the ground. As technology developed further, and resolutions improved, remote sensing has advanced to also detect and identify heat signatures of planted crops and animals. Similarly, moving beyond sonar, ocean temperature maps are used to show upwelling and chlorophyll distribution to identify coastal productive zones, use side-looking airborne radar to detect shoals of surface swimming fish, etc.

8.3.2 Geographical Information System (GIS)

GIS is usually a set of computer tools and is a unique platform that allows one to work with multiple data that are tied to a spatially mapped location or area on earth. This allows for multiple data of varied detail to be graphically depicted on a map and thus providing visual and other indicators to ease associated decision making. Advanced computer technology can now support provision of query based information and allow for the rapid computer analysis, for use at individual farms and across large territories. GIS tools and analytics can accurately depict the collection of data on, crop acreage, production, crop health, disease and also maintain geo-database of farmers.

8.3.3 Crop Cutting Experiment (CCE)

In the absence of modern technology based solutions, Crop Cutting Experiments (CCEs) are used to estimate the yield of a crop. This is essentially a statistical survey method, used to support crop insurance schemes that are area and yield based. In many countries crop yield estimation is based on field reports using conventional techniques for data collection. Satellite images are used where possible, for identification of the sampling plots. Nevertheless, CCE methods are time-consuming and costly, prone to large errors due to incomplete ground observations, and often lead to poor crop yield and crop area estimations. Recent advancements in technology have made it possible to rely on high-resolution satellite data sets on periodic basis. Reliable and timely information on crop area is extracted very efficiently, which can be correlated with yield data.

8.3.4 Crop Health Monitoring

The crop condition information at early stages in the crop growing season is more important than acquiring the exact production after harvest time. The district-wise crop health condition assessment is possible for major crops, viz, cotton, groundnut, paddy, wheat, potato, rapeseed, gram, tobacco, cumin, jowar, etc., using vegetation index computed from multi-date atmospherically corrected high-resolution satellite data.

8.3.5 Drones in Agriculture

Today, agriculture is one of the major adopters of drones in management. Drones can be ground-based and aerial-based and are increasingly used for field analysis to monitor crop health, irrigation, pesticide management, planting, etc. Drones can be used for conducting aerial surveys at regular interval to study the difference in land use, crop loss assessment, crop health imaging, and integrated GIS mapping. The uses of drones for gathering valuable data via a series of sensors, multispectral, thermal, and visual, for use in analytics, mapping and surveying of agricultural land. Depending on the sensors deployed on a drone, various data can be captured, such as plant health indices, plant counting, plant height measurement, canopy cover mapping, field water poising mapping, scouting reports, stockpile measuring, chlorophyll measurement, nitrogen content in the crops, drainage mapping, weed pressure mapping, and strikes of locust swamp on crop plants so on.

8.3.6 Upcoming Technologies

Developing technologies such as Big Data Analytics, Internet of Things (IoT), Block Chain, Artificial Intelligence, Robotics & Sensors, etc. are inter-related and are used to optimise the decision making process, and the operating procedures of every sphere where they find application. These technologies are practices that are deeply inter-woven with computerised systems, complex digitised interactions and even self-learning models. Agriculture no longer drives other economic and social activities, as in the past, but is subject to and is expected to reflect the demand from the wider population. Though the agricultural system, directly impacts on quality of life of all individuals, even those in non-agricultural activities, it in turn, is expected to be led by the demands from its end-consumers. A physically inter-connected world has made agriculture a highly competitive production and marketing system. Nevertheless, agriculture still has a certain fuzzy logic built into its operations, as the factors that affect the system, have various degrees in how they manifest. The widening scope of agricultural activities, its continued subjectivity to uncontrollable environs, the large quantity of data it generates from dispersed locations, and the increasing need to have focused & specific deployment of agricultural sciences has made the agricultural system an important domain for use of aforesaid new technologies.

8.3.7 Big Data from Agriculture

Big data analytics provides the opportunity to systemise the large amount of widely dispersed data that is generated from agricultural and allied activities. Farms and farmers produce big data, which need interpretation using Information technology for transforming the agricultural value system. As automation use in on-field and off-field machinery increases, large data from sensor technology will also be available. Such sensors are already seen in irrigation system, temperature control equipment, soil monitoring equipment, etc. The physical measurement and monitoring mechanisms, deployed through mobile imaging, satellite imagery, drone patrolling. Big Data Analytics gets its penetration by adopting technologies viz., Social Computing, Internet of Things (IoT), Data Virtualization, Statistical Methods and Machine Learning, Data Science Methods and Tools, Data Mining Algorithms, Data Analytics Processes, Platforms, and Practices, and Information Visualization Tools and Dashboards. Big Data Analytics is still at an early development stage in India. However, government agricultural development schemes (spread across the entire agricultural value system), AGMARKNET/e-NAM, Soil Health Card, National Animal Disease Reporting System (NADRS), Kisan Call Centre Database, DBT schemes and others, are already driving the need for adoption of big data analytics in the agricultural sectors.

8.3.8 Internet of Things (IoT) in Agriculture

Internet of Things (IoT) basically means the internet inter-connectivity of all connectable things. The things, include people and devices, i.e. anything that outputs or can use a 'yes/no' or 'on/off' signal. It connects the otherwise disconnected and disjointed machines. Further, IoT is a giant network of things which results in a connected relation between people-people, people-devices and devices-devices. This allows the things to digitally interact to trigger action or decisions, on the basis of usable information or pre-determined 'flags'. As a technology, IoT takes forward the networking of traditional devices like desktops, phones and tablets, to a wider range of everyday things like appliances, sensors, vending machines, automobiles, etc. IoT facilitates the remote access to active information. For example, a farmer visiting the local market, can access his in-field soil sensor to decide on what fertilizer to buy, or remotely access the health parameters of his livestock. Moving ahead, the Web of Things (WoT), is a refinement of the Internet of Things by integrating smart things not only into the internet (network), but into the Web Architecture (application), for ease of use and bring a higher level of maturity for scalability and sustainability.

8.3.9 Artificial Intelligence

Artificial Intelligence (AI) takes automation to another level, by incorporating analysis and learning on the basis of past and current data. AI supports in decision making, provided through machine and digital learning processes. Human intelligence can take long to assimilate, understand and react to all the complex variables that comprise the uncertainties that agriculture is subject to. AI can help make better sense of the inherent fuzzy data and rapidly put out answers from extremely complex inputs. Farmers can benefit not only from the direct on-farm applications of AI, but also from its use in the development of improved seeds, crop protection, and fertility products. Besides the unpredictable biological and weather related processes during cultivation, agriculture is also dependent on variables from multiple market situations. In this complex situation, applications based on AI require large amounts of data to properly train the algorithms. AI deployment is a natural corollary of large data warehousing (big data) and automation. Cognitive technologies allow analysing and correlating information about weather, type of seeds, types of soil or infestations in a certain area, probability of diseases, data about what worked best, year to year outcomes, market trends, prices or consumer needs; and in the final analysis facilitate farmers to make decisions to maximise on crops and livestock output. It may also be appreciated, that two of the important Public Extension Service Centres, are Krishi Vigyan Kendras (KVKs) and Agricultural Technology Management Agencies (ATMA). Both these are well positioned to be the nerve centres for AI applications, and for knowledge diffusion among India's vast farming community. The AI technology can be used in following areas:

- **Crop and Soil Management:** With the development of AI technology, it is easier to keep a track and predict the right time for planting, irrigation, and harvesting. The advanced sensors and technologies, make the entire task of crop and soil management uncomplicated for the farmers.
- **Pest Attack prediction:** Common pest attacks, such as jassids, thrips, whitefly, and aphids can pose serious damage to crops and impact crop yield. Use of AI and machine learning can indicate in advance, the risk of pest attack. This empowers the farmers to plan in advance, and benefit from reduced crop loss due to pests and as a result realise higher farm returns as presently the drone has been used to spray the pesticides against the swamp of locust in Rajasthan, Haryana, Uttar Pradesh and adjoining areas
- **Image recognition:** Artificial Intelligence can also be used for recognizing weeds and assessing plant health. Use of AI can differentiate between plants and weeds by leveraging big data, and actively sprays weedicides on the weeds, but ignores the crop plants.
- **Robotics applications:** As the farmers are automating their operation, robots and drones have become an integral part of the agriculture farms and are assisting farmland owners to improve the yield and product quality while addressing the increasing supply needs. With fragmented farms whose locations may be dispersed in different directions, it's almost impossible for a single farmer to go around and tend to all parcels of his farm that need watering or a measurement. This is where robots come into play for they can be pressed into service as per local needs.
- **Animal Husbandry:** Animal husbandry is an integral branch of agriculture concerned with the care and management of the livestock. It deals with all the tools and technologies involved in managing and ensuring optimum health of farm animals, including genetic qualities and behaviour. Generating and leveraging useful information through AI will help farmers to manage their livestock efficiently with minimum supervision. With AI enabled smart sensors, the automated milking units can analyse the milk quality and flag for abnormalities in the product.

8.3.10 Blockchain Technology for Agricultural Value System

Blockchain (database system), created by an unknown person or persons (named Satoshi Nakamoto), that maintains and shares a transparent immutable record of the history of the transactions. In the blockchain, a single ledger of records is shared with the transacting parties, where each must give consensus before another transaction is added, and once recorded, the transaction cannot be altered. In any supply chain, the blockchain using parties could include the producers, retailers, logistics providers, and regulators. Major food companies have commenced using Blockchain technology to transform their supply chain. In the agricultural sector, Blockchain technology can also be used to record inter-linked field practices such as INM/IPM (Integrated Nutrient Management/Integrated Pest Management), confirm good agricultural practices, validate resource use efficiencies, build traceability for the produce from farm to fork, prevent price extortion and delayed payments. Blockchain adoption requires access to reliable internet connection. The initiatives to bring last mile connectivity through wi-fi hotspots and mobile data in rural India, will enable such technologies. The proposed COOPNET (Cooperative Informatics Network) networking more than 100,000 Primary Agricultural Cooperative Societies (PACS) can be operationalised with such distributed cryptoledger (Blockchain) systems. The way forward is to establish a "Blockchain Technology based Testbed" for Agricultural Value System.

8.3.11 Robots and Sensors in Agriculture

Agricultural sector remains labour intensive and is a source of employment (through drudgery borne and unproductive, making replacement by machines a worth-while pursuit) to a large section of societies across the world. However, there are specialised areas where robotics has already come into use in agriculture. While robotics and sensors are a physical equipment, they use and provide inputs as digital signals and as a system are also considered part of digital technologies. Robotics also help in automation of some tasks and can also free the individual farmer to prioritise and take on other works for added gains. Robotics and automation are commonly used in nurseries for seeding, potting and care of the plants. Solutions are also in use in dairy facilities as in case of feeding and milking machines. Similar examples are seen in the fisheries sector, where automated feeders and pond aeration systems are used. Other reported on-farm uses are machines that recognise patterns and undertake targeted spraying of pesticide and fertilizer, the precision allowing to limit the application to individual plants. This functions much like face recognition in smart phones, where a data set of patterns triggers a precise reaction. Drones are unmanned flying machines, and make effective aerial sensors, used to control robotic machines for planting, seeding and harvesting large fields, besides other applications. The aerial view provided by drones, which can be across the digital spectrum, is digitally transmitted to computer systems, linked with geo-spatial signals and correlated and coordinated for various purposes. In green houses, much like rain sensing windshield wipers in automobiles, rain and light sensing robotic arms can automatically retract or cover the roof as per need. Similarly, hi-tech sorting and grading machines in modern pack-houses, sift and assay produce on the basis of optical and physical sensors, automatically package, label and move the boxes to next stage of handling.

Automatic fork-lifts, pallet put-away and picking arms, and many such uses are seen in modern cold stores and warehouses. Robotic and semi-robotic equipment are also used in poultry harvesting factories and abattoirs, besides in beverage factories, and the like. Robotics ease the physical handling of activities and large loads, doing it faster than humans can. The first level of smartness is derived from sensors, to start or stop an automation on its pre-determined set of actions. Sensors and robotics go hand in hand, not only to actuate action, but also to monitor and stop an activity, at levels that include safety needs.

8.4 ICT Based Support for Farmers

8.4.1 Websites/Portals

In order to meet the information needs of the farmer, Ministry of Agriculture and Farmers' Welfare has developed different websites and web portals that allow farmers to access the information using Internet. Information on Market Price, Soil Health Card, Crop Insurance, Government schemes etc. is available to farmers through these websites. These websites also aim at enhancing communication between the research institutions and the farmers. They have also helped improve communication and knowledge sharing between researchers and subject-matter experts. Farmers' Portal, Agmarknet, Soil Health Card Portal, eNam, Crop Insurance etc. are some of the examples of web portals developed for farmers.

8.4.2 Use of Mobile Apps

Diffusing agricultural related information to farmers spread across the vast geography is made easier by proliferation of mobile phones. Today, mobile apps and services are being designed and released in different parts of the world. Mobile apps help to fulfil the larger objective of farmers' empowerment and facilitate in extension services which can address global food security, agriculture growth and farmers' welfare. The use of mobile applications providing price information to farmers can reduce market distortions and help farmers to plan production processes. For example, the M-Farm application in Kenya led to farmers changing their cropping patterns and some reported receiving higher prices at market as a result (Baumüller, 2015).

Some illustrations of mobile apps developed for farmers are:

- **Kisan Suvidha** mobile app provides information on five critical parameters—weather, input dealers, market price, plant protection and expert advisories. An additional tab directly connects the farmer with the Kisan Call Centre (KCC) where agriculture experts answer their queries. Unique features like extreme weather alerts and market prices of commodity in nearest market and the maximum price in state as well as India have been added to empower farmers in the best possible manner.
- **Pusa Krishi** app helps farmers to get information about latest technologies developed in research labs. This app is actually transferring the technologies from "Lab to Land". Agrimarket mobile App can be used to get the market price of crops in market within 50km of the devices location. This app automatically captures the location of person using mobile GPS and fetches the market price. Crop Insurance mobile app can be used to calculate the Insurance Premium for notified crops based on area, coverage amount and loan amount in case of loanee farmer.
- **CCE Agri** is a mobile app used for data collection and data monitoring in rural areas. Data of crop cutting experiments (CCEs) is digitized using this mobile app which definitely removes chances of human error and reduces the time in data collation. This app significantly improves data speed (from harvesting to insurance loss estimation) and biggest gain is data quality. Geo-tagging ensures field visit, photos mitigate the manipulation risk and data transfer greatly improves data consolidation/ analysis which eventually results in quick claim settlement. In rural areas, there are challenges on account of absence of or poor connectivity. Hence, this (CCE Agri) app has been designed in such a way, that data can be collected without internet connection and as and when internet is available, data can be pushed to the server.
- **EMA-i** is an early warning app developed by FAO to facilitate quality and real time livestock disease reporting captured by animal health workers in the field. EMA-i is integrated in the FAO's Global Animal Disease Information System (EMPRES-i) where data are safely stored and used by countries. EMA-i is easily adaptable to countries existing livestock disease reporting system. By supporting surveillance and real time reporting capacities at country level and improving communication between stakeholders, EMA-i contributes to enhance early warning and response to animal disease occurrence with high impact to food security

and livelihood. EMA-i is currently used in six countries in Africa (Cote d'Ivoire, Ghana, Guinea, Lesotho, Tanzania and Zimbabwe). (FAO, 2019).

8.4.3 Use of basic Mobile Telephony

Mobile telephony has transformed the tenor of peoples' lives. In India, increased penetration of mobile handsets, large number of potential users, increased spread of communication, and low cost of usage are leading to growth of large number of mobile based information delivery models for the agricultural sector. A few of the modes used to meet the information needs of the farmer are **SMS, IVRS, OBD, USSD etc**.

8.4.4 Kisan Call Centres (KCCs)

Kisan Call Centres (KCCs) were launched by the Ministry of Agriculture and Farmers' Welfare in 2004 to bridge the gap between farmers and the technology assessment. This initiative was aimed at answering farmer's queries on a telephone call in their own language / dialect. Presently, KCC services are managed from fourteen locations. Kisan call centers provide services to farmers where they can directly interact with the executives for their queries. They are provided information through mobile phones with regard to suitable techniques needed to maintain the fertility of the soil to increase production is showing positive results. Information communication technology can change the face of agriculture sector by raising crop productivity and profitability per unit area and resources. The mobile connectivity in terms of service users is increasing at fast rate. IAMAI study showed 80% people in rural areas use teledensity for communication, 67% for online services, 65% for e-commerce and 60% for social networking (Upadhyay, 2019). All KCC locations are accessible by dialling a single nation-wide toll free number 1800-180-1551 through landline as well as mobile numbers of all telecom networks from 6 am to 10 pm. on all 7 days a week including holidays. KCC enables farmers to engage in direct discussions with the subject matter experts who are able to analyse the problem effectively and provide the solution directly. KCC uses a backend data support system, which is inbuilt into the overall MIS (Management Information System). The MIS software captures callers' details and specifications of the query which helps in analysing area-wise and crop wise details within a time space framework and provides preventive, advance action solutions. The long standing demand for scaling up the KCCs to 35, so that every state/UT has its own and enable language compliant and location specific knowledge sharing with the farmers.

8.4.5 National e-Governance Plan – Agriculture (NeGP-A)

In agriculture, availability of real time information at the right time is the major miss. Lack of information at proper time causes a huge loss to farmers, proving the adage, 'information is knowledge and knowledge is power'. NeGP-A aims to bridge this gap in communication by using technology. It provides an integrated approach to the delivery of services to the farming community using ICT. Under NeGP-A, around 60 online services have been developed over the last few years and launched to provide ease of access and timely information to farmers. Some services have been developed for monitoring of schemes, so that quick analysis and reporting can be done.

8.4.6 DINO Agrobot for Agriculture and Viticulture: The Naïo Technologies (Develop and markets machines for Agriculture & Viticulture) team developed agricultural robot to improve working conditions and profitability for farmers. To help farmers tackle the increasing regulations on phytosanitary products, the growing concerns with pesticides, and the lack of workers in the agricultural sector, Dino provides a new and effective solution. The Dino weeding robot allows vegetable farmers to manage crop weeding with a high level of precision, while helping them save time all through the season. Dino is highly effective to weed vegetables that are grown in the field, both in raised vegetable beds and in rows, such as lettuce, carrots, onions, etc.

8.4.7 MyCrop complete Farm and Farmer Management System: MyCrop a technology-enabled initiative for farmers, which empowers them through delivering information, expertise and resources, to increase productivity and profitability, hence improving standard of living. It is a collaborative platform that strives to combine cutting edge technology (Big Data, machine learning, smartphones/tablets, etc.), innovative business model (agriculture platform as a service), and focused human efforts (agriculture insights, products, and services) to serve smallholder farmers. MyCrop facilitates farmers in taking and executing optimum decisions by providing geomapping, crop planning, individual farm plans and farm automation customized for each farmer based on weather, soil, pest and crop data on an almost real-time basis. MyCrop is a sustainable data-driven, scalable, intelligent, self-learning, real-time collaborative Agrifood system, which serves as a farm as well as farmer management solution, predictive analytics and monitoring tool, decision support system and agriculture (buy/sales side) e-commerce platform.

There are various options for out scaling innovations in natural resource management. Innovation is an amalgamation of technology, local adaptation, social inclusivity and access by end-users. It is important to understand the big challenges associated with 'half innovations' and the successful conversion of half innovations into full innovations based on local needs. Major requirements for out scaling NRM innovations include long-term investment, portfolio of policies and practices, patience, capacity, innovation- led business models and robust *ex ante* analysis of return on investment. Scientific social responsibility/ science-corporate social responsibility needs to be given due importance. Since these NRM-based innovations generate a lot of social and environmental good, there is an urgent need for greater public investment in their promotion and use (Jat, 2017). Significant efforts have also been made to develop location-specific CSA practices through large national and international initiatives and programmes like the CGIAR research programme on Climate Change, Agriculture and Food Security (CCAFS), and the National Innovation on Climate Resilient Agriculture (NICRA) of the Indian Council of Agricultural Research (ICAR). The CSA technologies and learning from such programmes need to be replicated through enabling policies and increased investments. Efforts need to be made to capture farmer-led innovations on climate-smart agricultural practices and blend them with modern science. The learning through community-based approaches in climate- smart villages (CSVs) are good examples of blending science and society for participatory learning and building evidence on CSA (Jat, 2017), which needs to be replicated over large areas. For addressing issues of resource fatigue and bridging existing yield gaps, the recommendation domain of the best-bet CSA practices, resource mapping and characterization using new tools and techniques like remote sensing and GIS would help considerably. Documenting success stories of potential climate-smart technologies like stress-tolerant genotypes/hybrids, better feeding management of locally well-adapted livestock breeds, CA, laser land-levelling, micro-irrigation systems, use of customized fertilizer nutrients and replicating them in similar ecologies, production systems and farm conditions, through functional regional networks, should receive high priority. Digital Agriculture is no more a fantasy. A reality that has started to gain fractions in Indian soils, digital technologies such as Artificial Intelligence, Cloud Machine, Heavy Satellite Imaging and Advanced analytics are empowering small holder farmers to increase their income through high crop yield, better health input management and effective market price connect. Artificial Intelligence and big data are going to be a game changers in the agriculture sector, and the government is aiming to collate about 80 percent of such data beyond of 2020. The data will help in framing the right policy and converges some project in order to achieve the targeted development of farmers and the overall sector. Artificial Intelligence can be used in multiple domains of agriculture. Indian agriculture has been mostly traditional and the farmers have relied upon their perfectly honed agriculture wisdom in raising crops and protecting them. However, the challenges have broadened. The today and the future can no longer be dictated by individual farmers' cognitive abilities. Artificial Intelligence may help Indian farmers to choose the right crop and minimize the risks and raise farm incomes to decent levels (Khan, 2020).

8.5 Possible Areas of Smart Devices and Applications in Agriculture

The Department has communicated (DAC&FW, 2016) on the formation of the Working Group on SMART Village and Agriculture, and quotes areas related to Agriculture, where smart devices and applications can be useful, are as follows:

- Advise on requirement of water in the field in case of dryness increasing.
- Controlling the quantity, direction etc. in micro-sprinklers.
- Digital view of field from home /remote location.
- Environment control and weather sensors in poultry farms.
- Identification of Fish Density Zones (FDZs) in the Sea and informing and guiding fishermen about FDZs, helping them thereby with enhanced fish catch.
- Information on real time moisture and temperature variations in grain storage to undertake corrective measures to prevent pest infestation.
- IR sensors for monitoring growth and development in urban horticulture.
- Maintaining temperature/humidity for perishable goods.
- Monitoring body temperature of animals to ensure proper health and information on food quantity consumed from their food drum beds.
- Monitoring of incidence of insects and pests using sensors with various types of traps.
- Monitoring the soil density and nutrition values after a particular crop.
- Ontology based CDN Systems to assist the sensors for capturing data and analysing with standard data sets for a specific crop domain dynamically.
- Precision Livestock Farming (PLF) sensors for monitoring and early detection of reproduction events and health disorders in animals.
- Provide advice on crops to be sown based on nitrogen, phosphorous, and potassium (NPK) content and environment values and comparing with standard databases.

- Provide alert on any crop damage in case animals destroy crops.
- Provide details of dimensions of field and classification of crop planting by analysing the total field level coordinates.
- Provide details of water level, humidity/moisture content and acidity/pH in the soil.
- Provide information on protected cultivation in greenhouses.
- Provide inside and outside temperatures of the field in respect of crop as per its standards.
- Provide soil composition of soil components in the field.
- Provide temperature levels and also share in case of possibility of severe weather change by connecting the filed level device with district weather departments.
- Provide the status of growth of crop in terms of expected growth and current growth by using various parameters.
- Quality evaluation of honey bees and honey bed growth using IR (infra-red radiation) or UV (ultra violet) radiation analysis.
- Recommendations on the basis of crop growth, stage-wise nutrient content and required nutrients.
- Setting up of Digital Scare Crows, as well as monitor any field level intrusions.
- Temperature regulator in fisheries to produce different varieties of fishes in different zones like greenhouses.
- Use of bio-sensors matrix and analysing the weeds and insects in the field.
- Wireless communication to villages and online Knowledge-bases.

Science plays an important role in reducing costs, both for consumer and producer, makes agriculture sustainable by targeting improvements in resource use efficiency and countering external debilitating forces, both biological and physical.

8.6 Digitalisation of Villages

A Digital Village is often seen as an idea that can help remove the Digital Divide, enabling development to reach underdeveloped regions, and the country to leapfrog. In physical terms, a 'Digital Village' generally refers to a village that has voice as well as data connectivity. Though not specified, the assumption is that both are of sufficiently high quality, with bandwidth available for most commonly used Internet applications to work on computers, tablets and mobile phones.

Sh Piyush Goyal, Minister for Commerce and Industries in 2020 said that the country's digital infrastructure and economy by 2030 will be built on the work done in digitisation of government processes and private transactions. As part of this move, the government plans to make one lakh villages Digital Villages over the next five years with the help of Common Service Centres (CSCs), a public-private initiative that offers digital services to villagers.

8.6.1 Ongoing Efforts for Sustainable Development of Village

The demand of rural India today is sustainable growth and development. Expanded reach of the Government – both spatial and demographic – is the corner stone of e-Governance. In 1990s, when digital technologies for village level development were rolled out, it only had database technology and computer technology that merely facilitated management information system (MIS) reports. India needs an economic movement that starts in villages, and not one that tends to bypass them. There had been many efforts to establish "Village level Database" for micro level planning and decision support, and "Village level Knowledge Management System" for checking farmers' distress (e.g. Information Village Project of IDRC/MSSRF Chennai, Village Resources Centre of ISRO, Village Knowledge Centre of CAPART, and Village Knowledge Centre of Union Bank of India etc).

Village Knowledge Centres (VKCs) were envisaged as information dissemination centres providing the farmers instant access to latest information/ knowledge available in the field of agriculture, starting from crop production to marketing. The National Alliance for Mission 2007 Initiative had received support from the United Nations Development Programme (UNDP), the International Development Research Centre (IDRC) and the Canadian International Development Agency (CIDA), the Swiss Agency for Development and Cooperation (SDC), the United Kingdom's Department for International Development, the World Bank, the International Crops Research Institute for the Semi-Arid Tropics (ICRISAT), the United Nations Educational, Scientific and Cultural Organisation (UNESCO), the World Health Organisation (WHO), the Food and Agriculture Organisation (FAO), the World Food Programme (WFP), the International Fund for Agricultural Development (IFAD), the McArthur Foundation, the Jhai Foundation, and the Global Knowledge Partnership. This National Alliance, then, included 22 Government Organisations including the Ministry of Information Technology, the Ministry of Panchayati Raj, the Telecom Regulatory Authority of India

(TRAI), and Bharat Sanchar Nigam Limited (BSNL); 94 Civil Society Organisations; and 34 Private Sector Information and Communication Technology (ICT) leaders such as NASSCOM, TCS, HCL, and Microsoft. Besides, 18 Academic Institutions such as the Indian Institutes of Technology, and the Indira Gandhi National Open University; and 10 financial institutions such as the National Bank for Agriculture and Rural Development (NABARD) and the State Bank of India (SBI).

Two initiatives, namely, GRID (Grassroots Level Informatics Development Programme) and SMART Village Project of National Informatics Centre (NIC) were envisaged, during 2007-12, to provide the benefits of Information and Communication Technology (ICT) directly to the communities at the grass roots level. In 2008, the Department of Science & Technology (DST), GOI, took keen interest in operationalising a systemic response to prevent distress such as farmers' suicides in suicide-prone / affected villages, through setting up of Village Knowledge Management System (VKMS). The PURA (Provision of Urban Amenities to Rural Areas) 2.0 was launched in 2012. Now in operation is the Shyama Prasad Mukherjee Rurban Mission (SPMRM) - National Rurban Mission 2016, a successor to PURA 2.0, to deliver integrated infrastructure cluster in the rural areas, including promotion of economic activities and skill development – 300 SMART Village Clusters within 3 years, based on Integrated Cluster Action Plan (ICAP) with 14 mandatory components as presented below:

8.6.2 Digital Technologies

Spatial Technologies, Artificial Intelligence (AI), Geographic Information System (GIS), Remote Sensing, Drones, Database Technology, Data Analytics, Blockchain Technology, Internet & Web, I-SMAC Technology etc. These are now available to facilitate village development linked to plans and programmes, based on its resources.

8.6.3 Digital India Programme

The Digital India Programme, launched on 20th August 2014, promises to transform India into a connected knowledge economy offering World-Class Services at the click of a mouse, and has been envisaged with the following 9 Pillars of Growth: - The earlier National e-Governance Programme (NeGP) has been re-categorised into e-Kranti (electronic delivery of services) and e-Governance (reforming Government through technology) Programmes.

8.6.3.1 Common Services Centre (CSC)

Common Services Centres (CSC) scheme is one of the mission mode projects under the Digital India Programme. CSCs are the access points for delivery of essential public utility services, social welfare schemes, healthcare, financial, education and agriculture services, apart from a host of B2C (Business to Customers) services to citizens in rural and remote areas of the country. It is a pan-India network catering to regional, geographic, linguistic and cultural diversity of the country, thus enabling the Government's mandate of a socially, financially and digitally inclusive society. The CSCs enable the three vision areas of the Digital India programme: Digital Infrastructure as Core Utility to Every Citizen; Governance and services on demand; and Digital empowerment of citizens. Among the others, CSCs also provide high quality and cost-effective video, voice and data content and services, in the areas of e-Governance and Agriculture Services (Agriculture, Horticulture, Sericulture, Animal Husbandry, Fisheries, and Veterinary). It offers web-enabled e-governance services in rural areas.

8.6.3.2 National Centre of Geo-Informatics (NCoG)

In view of the importance of geo-informatics in e-Governance, the National Centre of Geo-Informatics (NCoG) of the Ministry of Electronics and Information Technology, has been established to promote "geo-informatics" technology, through its "GIS based e-Governance Process". NCoG envisages to

- Develop collaboration with both private and public sectors.
- Develop programs and capacities related to human resources.
- Take up R&D in the area of geo-informatics.
- provide geospatial applications and solutions to all Governments – central, state and local governments,
- provide national platform for developing geo-informatics in the country,

8.6.3.3 Last Mile Connectivity

Last-Mile technology represents a major challenge due to high cost of providing high-speed and high-bandwidth services to individual subscribers in remote areas. Laying of wire and fibre optic cables is an expensive undertaking that can be environmentally demanding and requires high maintenance. Broadband wireless / wired networks viz., BharatNet, Cable TV Networks, RailTel, Electricity Lines, LoRaWAN, TV White Space Technology etc., will eventually be required to provide the solution to achieve "last mile connectivity" of Digital India Programme. The on-going Digital Network for Farmers (DNF) over the Broadband Wireless/Wired Network with APP such as Krishak Mithra Software (KMS) will establish the "last mile connectivity" to have farmers "digitally included" for ushering in "Digital Agriculture India" effectively.

8.7 Digital Village Project

Digital Village Project, among others, aims at the usage of Information & Communication Technologies (ICTs) for development and empowerment of communities (mostly disadvantaged communities). The initiative aims to empower communities (that have limited or no telecommunications access) through the use of mobile technologies, which will help contribute to long-term sustainable and economic development, through supply-chain modules. The model will work around the human resources at the village level as an individual, family and society; working out the linkages, identification of the right stake holders, analysis of the services, working out the methodology, digital enablement with digital connectivity with the right stake holders and necessary infrastructure at the village level. The project proposes to realize this, through the following:

- Introduce & promote Information and Communication Technologies that are cost effective and appropriate for use in rural areas, to enable rural villages digitally to access and benefit variety of services at the last-mile.
- Develop and implement a service model wherein the villagers, NGOs and the Government work as a cohesive unit in building, maintaining and delivering the information and knowledge base to facilitate development and empowerment of the community.
- Explore and strengthen avenues to make the service model self–sustainable at village level.
- Delivery of goods and services right from the request / registration at the village level to delivery at the doorsteps, through the use of agile methods for refining the processes every time in the Service Delivery Life Cycle (SDLC).
- Operationalise BOM (Build, Operate and Maintain) Model for incubation/deriving best practices for two years and thereafter, a realistic sustainable ROT (Re-model, Operate and Transfer) Model wherein all the failure entities and processes are removed, new innovation, technology updates, process optimization are introduced for five years.

The major focus areas, among others, will be:

- **e-Agriculture:** Crops, Livestock, Fisheries (inland and marine), Agro-forestry, Forestry (Minor Forest Produce), Water and Agriculture in areas dominated by tribal communities.
- **e-Education:** Literacy and technical skills; Digital Learning etc.

- **e-Governance:** Information on entitlements and on methods of accessing the entitlements (e.g. bank credit, inputs, etc).
- **e-Judiciary:** knowledge of legal systems and processes.
- **e-Health:** Disease prevention, detection and cure; nutrition with particular reference to maternal and infant (0-2 years) nutrition.
- **e-Environment:** Conservation and enhancement of natural resources, with specific attention to land care, water conservation and sustainable use, conservation of flora and fauna and management of common property resources.
- **e-Disaster Management:** Methods to secure investments, cope with disaster and survival in case of floods, cyclones and rare events.
- **e-Livelihoods:** Opportunities for on-farm and non-farm employment, micro-enterprises supported by micro-credit, new skills and training in agro-processing and agri-business.
- **e-Commerce:** Producer-oriented marketing, quality management, matching production with demand.
- **e-Traditional Knowledge** and practices.

8.8 Natural Resource Management System for Reduction of Vulnerability

Natural Resources Management (NRM) is closely associated with (a) farm, (b) socio-economic environment, and (c) bio-physical environment situations and the outcomes that need to be ensured are:

- That, value of output exceeds value of inputs
- That, inputs do not degrade resource base
- That, resource degradation does not exceed resource productivity
- That, system productivity does not exceed resource productivity

For scientific utilisation of natural resources base, it is considered that product of interaction of rain with land, in other words, watershed is an ideal geographical unit. Each watershed contains a complex mixture of: soil types, landscapes, climatic regimes, land use characteristics, and agricultural systems, and can be subdivided into agro-eco-regions (AER) having similar soil types, landscapes, climatic regimes, crop and animal productivity, and hydrologic characteristics.

8.9 Achievement in 2020, e-GramSwaraj Portal for Panchayat Raj

Prime Minister Narendra Modi has launched a new e-GramSwaraj portal for Panchayat Raj system in the country. The portal has been launched along with Swamitva Scheme for the panchayat raj. e-GramSwaraj aims to bring in better transparency in the decentralized planning, progress reporting and work-based accounting. According to the official statement, the portal is going to bring in transparency in the Indian panchayat system. At the same time, the scheme has been designed to end the property disputes, by mapping village properties through the use of drones. e-GramSwaraj (user friendly web-based portal) is a step towards complete digitization of villages and will maintain all records of all developmental activities of villages and will be accessible on mobile phones.

8.10 Challenges to Connect Marginalized and Remote Communities

A well-developed digital infrastructure, especially in rural areas, is a precondition for digital agriculture and food systems. Although advances in technology and regulatory reform have improved access to ICT for people around the world, there still exists a digital divide. Just as a certain technology (e.g. dial-up Internet) becomes available across income levels, a new technology (e.g. broadband) appears, leaving users in developing countries 'playing catch up'. Although mobile-cellar subscriptions in the last five years were driven by countries in Africa and Asia and the Pacific, many people still do not own or use a mobile phone and the distribution of ownership is unequal. Access to web-enabled smartphones and fast 3G or 4G internet connections remains particularly limited in rural areas. There will need to be work to address this disparity and to facilitate smartphone ownership and use in areas where it is currently lacking. Both literacy and education levels also remain particularly low for rural populations in developing countries and LDCs which presents a barrier to the use of digital technologies. Youth unemployment rates are often higher than the country average and this is especially the case in rural areas. Increasingly, employers want employees who are adept at using technology. A lack of e-literacy and digital skills in rural areas means these populations will fall behind in the modern labour market. There is a need for school curricula to incorporate digital subjects, for improved knowledge and skills among teachers and for increased availability of digital technologies in classrooms. To unlock the full potential of digital agriculture transformation, governments need to create an enabling regulatory environment. Designing and managing digital government programmes requires a high level of administrative capacity which is beyond the capabilities of some countries, particularly LDCs and developing countries. Addressing the digital divide must be made a policy priority and governments should make the socioeconomic case for digitalization of smallholder farming both to the farmers, and to potential private sector investors and start-up businesses. There will need to be significant capacity building among governments in developing countries and LDCs to facilitate this change in policy and regulation. There is increased interest in data-enabled farming and related services and many new entrants from the technology industry and start-ups. Vast data collection will drive the use of machine learning and AI and new models will need to be developed to make the data useful. So far, the information gathered is often insufficient to inform the comprehensive solutions and partnerships needed to transform smallholder farming into viable, sustainable digital businesses. There also need to be decisions about the ownership and use of data manufacturers collect data from their devices and have the opportunity to exploit them, but farmers are often reluctant to share their data without receiving something in return. Strategies for digital agricultural transformation in developing countries must combine IT infrastructure with social, organisational and policy change (FAO, 2019).

8.11 Drivers and Demands for Unlocking Digital Agriculture Transformation

Access to the internet remains the most critical component for unlocking the possibilities of new technologies. Across the globe, smartphones dominate in terms of time spent online and could be a game changer in the agrifood sector in LDCs and developing countries. They create opportunities to access information and services through mobile applications, online videos and social media. Sites like Facebook, Twitter and YouTube present a cost-effective means of communication with, and among, smallholder farmers and other key agricultural stakeholders such as extension officers, agro-dealers, retailers, agricultural researchers and policy makers (FAO, 2019). Falling handset prices, increasing internet coverage and the growing youth population create significant opportunities for the use of mobile phones in agricultural areas. However, internet provision and smartphone ownership remain lower in developing countries, and particularly in rural areas, and there needs to be more research into the use of mobile internet and social media in rural communities.

Additionally, not all farmers are quick to adopt ICT. Many lack the necessary knowledge to request or use services, especially as ICT applications in the agrifood sector are relatively new and many e-services are still being developed. It is critical that technologies are properly targeted; if they do not provide the information that farmers need, they will not be adopted. Digital skills and e-literacy remain a significant constraint to the use of new technologies and are particularly lacking in rural areas, especially in developing countries. The diversity of available digital technologies and a lack of standardisation also present a barrier to adoption. The choice of which technology to use is complex and there is a lack of advisory services to support farmers in these decisions. Education and supporting services must be improved to support the adoption of digital technologies. Digital technology is already changing the dynamics of the agrifood sector but the process has so far not been systematic. Realising the full potential of digital farming will require collaboration of all players in the agricultural value chain. There is a need for a clear overview on the part of actors working in agrifood and digital products – including private sector, governments and other agencies – on how to exploit the opportunities of digital agriculture. Farmers have a key role to play and digital technologies provide new opportunities for them to collaborate and innovate. There is also a growing group in the farming sector who have university degrees and specialisations in science and technology subjects. They are often skilled in experimentation and innovative thinking. Youth in the agrifood sector are also often entrepreneurial and willing to take calculated risks to pursue new enterprises.

Digital transformation will provide access to finance through exposure and awareness due to digitization, forecasts on climate change enables right decisions, accessibility of farm equipment and new technology, inputs for better soil fertility and soil structure, access to markets, access to information, small holdings utilization and enables predictive analysis. It requires planning, capacity building, identification of right stakeholders, mechanisms for governance and monitoring and provide buyers and sellers one platform. This technology platform will reduce costs, improve productivity and quality, improve prices, reduce risks and create sustainable ecosystem. Digital technologies offer the potential to achieve the necessary conditions for scale, with distributed low cost and customized delivery, creating a unique opportunity for private enterprise and innovation to thrive. High and inclusive growth can be well promoted with digitization. For India, at a time when national, regional and international research institutes have already developed technologies, farmers need motivation and encouragement to adopt these proven yield-enhancing, cost-efficient and environment-friendly technologies. Finally digitization will change the scene of Indian agriculture in future and guarantee higher income to farmers and reduce distress. The digitalization of agriculture will cause a significant shift in farming and food production over the coming years. Potential environmental, economical and social benefits are significant, but there are also associated challenges. Disparities in access to digital technologies and services mean there is a risk of a digital divide. Smallholder famers and others in rural areas are particularly at risk of being left behind, not only in terms of e-literacy and access to digital resources but also in terms of productivity and aspects of economical and social integration. Simply introducing technologies is not enough to generate results. Social, economic and policy systems will need to provide the basic conditions and enablers for digital transformation. The "Law of Disruption" (Downes, 2009) states that technology changes exponentially, but economic and social systems change progressively and have trouble keeping up. Work is especially needed to ensure the necessary conditions for digital transformation are created in rural areas.

References

Baumüller, H. (2015). EJISDC. (68) 6:1-16.

Bergvinson David (2017). Business Today retrieved from https://m,businesstoday.in

Downes, L. (2009). The Laws of Disruption: Harnessing the New Forces that Govern Life and Business in the Digital Age. Basic Books.

FAO (2019). Digital Technologies in Agriculture and Rural areas Briefing Paper by Nikola M. Trendov, Samuel Varas, and Meng Zeng Food and Agriculture Organization of the United Nations Rome.

Ganguly, S.; Patra , S.P. (2017). International Journal of Advance Research, Ideas and Innovations in Technology pp. 403-410.

http://www.mycrop.tech

https://www.naio-technologies.com/en/agricultural-equipment/large-scale-vegetableweeding-robot/

Jat, M.L. (2017). Climate smart agriculture in intensive cereal based systems: scalable evidence from the Indo-Gangetic plains. In: Belavadi, V.V., Nataraja Karaba, N. and Gangadharappa, N.R. (eds) Agriculture under Climate Change: Threats, Strategies and Policies. Allied Publishers Pvt Ltd, India, pp. 147–154.

Jat, M.L. (2017). Out scaling Natural Resource Management Innovations. Paper presented by Dr M.L. Jat, International Maize and Wheat Improvement Center (CIMMYT), NASC Complex, Pusa, New Delhi, during a dialogue on 'Incentives and strategies for scaling out innovations for smallholder farmers'. Organized by the Trust for Advancement of Agricultural Sciences, Pusa Campus, New Delhi, 30–31 October.

Paroda, R.S. (2018). Reorienting Indian Agriculture Challenges and Opportunities. Trust for Advancement of Agricultural Sciences (TAAS), IARI New Delhi, India.

UN DESA. (2018a). The 2018 Revision of World Urbanization Prospects. New York: UN DESA.

Upadhyay, M. (2019). International Journal of Innovative Social Science & Humanities Research 6(1): 5-9.

Annexure*

*Starts from next page

Annexure-I

State/Ut-Wise Biofertilizer Production (Carrier Base) (Mt) in India During 2016-17

State/UTs	Capacity	Azotobacter	Azospirillum	Rhizobium	PSB	KMB	ZSB	VAM	Acetobacter	NPK consortium	Others	Total
Andhra Pradesh	3860 .00	481.53	457.34	285.12	1217.10	189.42	13.9	628 .0	76 .0	16.1 0	11.4 0	3375.91
Arunachal Pradesh	120 .00	18.1 0	20.1	22.30	-	-	-	-	-	59.20	-	119.70
Assam	2350 .00	174.80	453.70	152.20	575.50	0.85	-	-	-	-	2.00	1359.05
Bihar	150 .00	26.00	27.00	23.00	31.00	-	-	-	-	-	-	107.00
Chhattisgarh	1815.00	52.174	0.84	129.28	515.52	-	-	257.26	-	-	-	955.07
Goa	822 .00	-	-	-	-	-	-	822.00	-	-	-	822.00
Gujarat	4370 .00	357.46	370.31	254.6	764.35	375.68	28.45	1646.8	112.17	-	-	3909.82
Haryana	11500	155.37	0	88.134	155.98	348.28	14.29	1438	160.00	0.59	-	2360.64
Himachal Pradesh	250 .00	0.304	1.1 0	0.312	0.56	1 .00	-	-	-	-	-	3.28
Jammu & Kashmir	0.8 0	-	-	-	-	-	-	-	-	-	-	-
Jharkhand	60 .00	4.158	-	8.502	5.892	-	-	-	-	-	-	18.552
Karnataka	46862	2760.31	4067.37	2241.75	18329.79	1489.29	1227.09	607.56	70.00	599.9	160.00	31553.06
Kerala	9046	238.12	481.3327	375.889	597.0685	1507.385	26.8	45.546	-	545.828	1175.90	4993.87
Madhya Pradesh	27800	553.817	11.201	707.639	1346.118	633.786	106.988	313.537	61.00	1066.42	808.50	5609.01
Maharashtra	16400	1703.82	589.09	1241.83	3205.639	649	198.42	384.22	195.54	127.72	28.337	8323.62
Manipur	50	5.00	5.00	5.00	5.00	5.00	-	-	-	-	-	25.00
Meghalaya	-	-	-	-	-	-	-	-	-	-	-	-
Mizoram	4.00	0.50	0.15	0.6	1.25	-	-	-	-	-	-	2.50
Nagaland	200	34.00	3.99	3.26	10.2	-	-	-	-	-	-	51.45

State/UTs	Capacity	Azotobacter	Azospirillum	Rhizobi-um	PSB	KMB	ZSB	VAM	Acetoba-cter	NPK consor-tium	Others	Total
Odisha	1440	106.06	23.311	177.66	209.25	-	-	-	-	-	-	516.28
Punjab	8975	398.373	190.07	4.03	949.245	510.594	1157.624	1155.388	0	1128.85	39.60	5533.77
Rajasthan	121.00	-	98.00	295.00	-	-	-	-	89.00	108.00	-	711.00
Sikkim	5.55	-	3.90	6.80	-	-	-	-	-	--	-	16.25
Tamil Nadu	67518.31	3451.908	5384.327	2412.143	3142.385	143.336	77.803	5623.88	267.35	1000	5924.83	27427.962
Telengana	-	-	-	-	-	-	-	-	-	-	-	-
Tripura	3000	382.30	389.10	-	382.10	-	-	-	-	-	-	1153.50
Uttarakhand	6450	84.53	26.40	26.10	95.86	392	391	424.90	-	1165.89	1114	3720.68
Uttar Pradesh	3116	688.37	16.7	326.55	1118.76	11.30	-	19.90	-	81.80	572.41	2835.79
West Bengal	13885	1092.7	495.54	545.34	688.32	34.98	-	-	-	-	338.30	3195.18
Andman & Nicobar	-	-	-	-	-	-	-	-	-	-	-	-
Chandigarh	-	-	-	-	-	-	-	-	-	-	-	-
Dadra & Nagar haveli	-	--	-	-	-	-	-	-	-	-	-	-
Daman & Diu	-	-	-	-	-	-	-	-	-	-	-	-
Lakshadweep	-	-	-	-	-	-	-	-	-	-	-	-
Delhi	1000	-	-	-	-	-	-	-	-	-	116.2	116.20
Pondicherry	2075	18.30	84.903	1.629	49.034	23.80	18.30	-	-	-	8	203.97
Grand total	233245.66	12788.004	13200.7747	9334.668	33395.9215	6315.701	3260.665	13366.991	1031.06	5900.298	10299.477	109020.11

Source: Annual Report 2016-2017, National Project on Organic Farming, NCOF

Annexure-II

State/UT-wise Biofertilizer Production (Liquid base) (Kl) in India during 2016-17

State/uts	Capacity	Azotobacter	Azospirillum	Rhizobium	PSB	KMB	ZSB	VAM	Acetobacter	NPK consortium	Others	Total
Andhra Pradesh	670	92.9	2.69	10.99	208.02	33.22	3.5	2.6	3.5	3.52	4.3	365.24
Arunachal Pradesh	-	-	-	-	-	-	-	-	-	-	-	-
Assam	1400	3	-	-	3	20	-	-	-	-	-	26
Bihar	-	-	-	-	-	-	-	-	-	-	-	-
Chhattisgarh	800	0.5	1.48	0.19	2.63	2.04	1.89	-	-	1.5	-	10.23
Goa	-	-	-	-	-	-	-	-	-	-	-	-
Gujarat	6575	472.51	289.59	188.74	572.51	338.6	26.45	-	86.33	668.04	215	2857.77
Haryana	1590	5.078	0.05	6.6	10.04	14.82	1.37	-	0.09	32.1	-	70.148
Himachal Pradesh	450	20	-	5	24.5	21	20	3.9	-	50.2	50.1	194.7
Jammu & Kashmir	-	-	-	-	-	-	-	-	-	-	-	-
Jharkhand	-	-	-	-	-	-	-	-	-	-	-	-
Karnataka	2589.9	51.946	419.783	90.569	191.987	102.627	48.385	-	-	85.796	2.35	993.443
Kerala	184.5	7.264	9.0781	14.7958	15.164	7.3124	-	6	-	-	-	59.6143
Madhya Pradesh	1745	27.772	10.36	25.29	131.869	9.774	9.838	-	7.31	2.39	13.5	238.103
Maharashtra	866	67.75	41.97	63.82	86.52	64.33	0	19.88	17.7	11.37	24.99	398.33
Manipur	-	-	-	-	-	-	-	-	-	-	-	-
Meghalaya	-	-	-	-	-	-	-	-	-	-	-	-
Mizoram	-	-	-	-	-	-	-	-	-	-	-	-
Nagaland	-	-	-	-	-	-	-	-	-	-	-	-
Odisha	1350	8.958	6.729	6.89	7.713	0.7	0.8	-	-	-	-	31.79

State/uts	Capacity	Azotobacter	Azospirillum	Rhizobium	PSB	KMB	ZSB	VAM	Acetoba-cter	NPK consortium	Others	Total
Punjab	1870	46.61	15.16	14.204	34.261	11.13	1.61	0.062	0.67	86.47	-	210.177
Rajasthan	-	-	-	-	-	-	-	-	-	-	-	-
Sikkim	-	-	-	-	-	-	-	-	-	-	-	-
Tamil Nadu	2278.5	18.896	171.371	55.132	514.902	46.407	12.424	3.3	7	8.54	37.32	875.292
Telengana	-	-	-	-	-	-	-	-	-	-	-	-
Tripura	-	-	-	-	-	-	-	-	-	-	-	-
Uttarakhand	1400	91.38	15.35	28.75	36.03	15.2	15.2	28.8	-	230.48	-	461.19
Uttar Pradesh	380	177.57	3.9	-	333.1	17.01	3.08	-	21.18	139.82	1.24	696.9
West Bengal	120	4.09	0.95	8.58	9.87	0.97	0.5	0.4	-	0.85	-	26.21
Andman & Nicobar	-	-	-	-	-	-	-	-	-	-	-	-
Chandigarh	-	-	-	-	-	-	-	-	-	-	-	-
Dadra & Nagar haveli	-	-	-	-	-	-	-	-	-	-	-	-
Daman & Diu	-	-	-	-	-	-	-	-	-	-	-	-
Lakshadweep	-	-	-	-	-	-	-	-	-	-	-	-
Delhi	-	-	-	-	-	-	-	-	-	-	-	-
Pondicherry	1100	1.194	1.393	0.33	2.42	2.6	0.73	-	1.26	0.92	0.35	11.197
Grand total	25368.9	1097.418	989.8541	519.8808	2184.536	707.7404	145.777	64.942	145.04	1321.996	349.15	7526.334

Source: Annual Report 2016-2017, National Project on Organic Farming, NCOF

Annexure-III

Zone- Wise Bio-Fertilizer Production in India (2011-12 to 2016-17)

	State Carrier based (MT)	2011-12	2012-13	2013-14	2014-15		2015-16		2016-17	
		Carrier based (MT)	Carrier based (MT)	Carrier based (MT)	Liquid based (KL)	Carrier based (MT)	Liquid based (KL)	Carrier based (MT)		Liquid based (KL)
South Zone										
1	A & N Islands	0	0	0	0	0	0	0	0	0
2	Andhra Pradesh	1126.35	1335.74	2714.22	2668.80	274.8560	3062.6	317.811	3375.91	365.24
3	Daman & Diu	0	0	0	0	0	0	0	0	0
4	Karnataka	5760.32	7683.72	9907.337	16462.6200	23.0561	23042.91	488.142	31553.06	993.443
5	Kerala	904.17	1045.64	3520.66	4916.9700	10.5096	4926.045	56.5751	4993.8692	59.6143
6	Lakshadweep	0	0	0	0	0	0	0	0	0
7	Pondicherry	509.45	621	516.98	560.95	1.4976	283.641	4.088	203.966	11.197
8	Tamil Nadu	3373.81	11575.7	14104.83	15373.29	11.3017	23721.2104	861.9535	27427.962	875.292
		11674.1	22261.8	30764.027	39982.63	321.221	55036.41	1728.57	67554.7672	2304.7863
West Zone										
1	Chhattisgarh	276.34	501.63	712.07	1024.68	9.620	954.371	9.38	955.074	10.23
2	Gujarat	2037.35	978.48	6411.434	3667.929	2800.500	3963.42	2873.317	3909.82	2857.77
3	Goa	0	370	66.26	802.520	0	820.52	0	822	0
4	Madhya Pradesh	2309.06	1408.08	4824.194	2637.990	119.216	2741.30775	131.033	5609.006	238.103
5	Maharashtra	8743.69	5897.91	6218.607	14847.397	324.767	7825.142	389.665	8323.616	398.33
6	Rajasthan	199.78	982	1315	599.898	0	680	0	711	0
7	D & N Haveli	0	0	0	0	0	0	0	0	0
		13566.220	10138.100	19547.565	23580.414	3254.103	16984.76	3403.395	20330.516	3504.433
North Zone										
1	Delhi	1617	0	396	104.500	0	106.2	0	116.2	0
2	Chandigarh	0	0	0	0	0	0	0	0	0

	State Carrier based (MT)	2011-12	2012-13	2013-14	2014-15		2015-16		2016-17	
		Carrier based (MT)	Carrier based (MT)	Carrier based (MT)	Liquid based (KL)	Carrier based (MT)	Liquid based (KL)	Carrier based (MT)		Liquid based (KL)
3	Haryana	914.41	5832.61	1146.483	872.955	46.489	1097.457	58.032	2360.644	70.148
4	H.P.	1.29	0	26.147	0.768	33.070	2.712	190.05	3.276	194.7
5	J & Kashmir	0	0	45.26	0	0	0	0	0	0
6	Punjab	692.22	2311.33	2124.852	6305.453	74.278	2197.197	149.581	5533.774	210.177
7	Uttar Pradesh	8695.08	1310.02	2682.221	4099.068	98.036	3053.115	223.34	2835.79	461.19
8	Uttarakhand	263.01	2758.21	5493.851	2129.952	208.034	3549.39	428.22	3720.68	696.9
		12183.010	12212.170	11914.814	13512.696	459.907	10006.07	1049.223	14570.364	1633.115
East Zone										
1	Bihar	75	52.4	52.4	64.90	0	97	0	107	0
2	Jharkhand	8.38	35.3	14.2	9.08	0	9.172	0	18.552	0
3	Odisha	590.12	407.1	1097.61	1074.46	4.70	467.634	13.701	516.281	31.79
4	West Bengal	603.2	1110	1682.7076	2061.83	14.63	2826.27	23.537	3195.18	26.21
		1276.7	1604.8	2846.9176	3210.27	19.33	3400.076	37.238	3837.013	58
North East Zone										
1	Arunachal Pradesh	0	0	59	59	0	3062.6	317.811	119.7	0
2	Assam	68.33	89	149	88	0	1315	22.5	1359.05	26
3	Manipur	0	0	0	0	0	0	0	25	0
4	Meghalaya	0	0	0	0	0	0	0	0	0
5	Mizoram	0	0	4	3.60	0	4.2	0	2.5	0
6	Nagaland	13	7.45	7.45	7.45	0	8.81	0	51.45	0
7	Sikkim	0	9.5	10.1	12.4	0	12.91	0	16.25	0
8	Tripura	1542.85	514	225	240	0	1143.07	0	1153.5	0
		1624.18	619.95	454.55	410.45	0	5546.59	340.311	2727.45	26
	Grand Total	40324.21	46836.82	65527.87	80696.45	4054.56	88029.30	6240.92	109020.11	7526.33

Source: Annual Report 2016-2017, National Project on Organic Farming, NCOF

Annexure-IV

State-wise Biofertilizer Production (Carrier + Liquid base) in India during 2016-17

State	Carrier (MT)	Liquid (KL)
Andhra Pradesh	3375.91	365.24
Arunachal Pradesh	119.70	-
Assam	1359.05	26.00
Bihar	107.00	-
Chhattisgarh	955.07	10.23
Goa	822.00	-
Gujarat	3909.82	2857.77
Haryana	2360.64	70.148
Himachal Pradesh	3.28	194.7
Jammu & Kashmir	-	-
Jharkhand	18.552	-
Karnataka	31553.06	993.443
Kerala	4993.87	59.6143
Madhya Pradesh	5609.01	238.103
Maharashtra	8323.62	398.33
Manipur	25.00	-
Meghalaya	-	-
Mizoram	2.50	-
Nagaland	51.45	-
Odisha	516.28	31.79
Punjab	5533.77	210.177
Rajasthan	711.00	-
Sikkim	16.25	-
Tamil Nadu	27427.962	875.292
Telengana	-	-
Tripura	1153.50	-
Uttarakhand	3720.68	461.19
Uttar Pradesh	2835.79	696.90
West Bengal	3195.18	26.21
Andman & Nicobar	-	-
Chandigarh	-	-
Dadra & Nagar haveli	-	-
Daman & Diu	-	-
Lakshadweep	-	-
Delhi	116.2	-
Pondicherry	203.97	11.197
Grand total	109020.114	7526.3343

Source: Annual Report 2016-17 National Project on Organic Farming, NCOF

Annexure V

Organic Exporters

Name and Address of Exporter	Name and Address of Exporter	Name and Address of Exporter	Name and Address of Exporter
Chamong Tea Exports (P) Ltd, 2, N.C. Dutta Sarani, Estate, 5th Floor, Unit-1 Kolkata 700 001 West Bengal 33-22203742/2243 4979 033 – 2243 7923 chamong@snonline.com	BBTC Export Operations Post Box No. 573 Subramanian Road Willingdon Island Cochin 682 003 Kerala 0484- 667539/666251/2 0484- 668321 ecotea@vsnl.com	L.T.Overseas Ltd. A-21, Green Park, Aurobindo Marg New Delhi – 110 016 Tel: 91-11-2685 9244, 26513450 91-11-26859344/ 26513450 Email: ltoltd@del2.vsnl.net.in	Balmer Lawrie & Co. Ltd P-43, hide Road Extension, Kolkata - 700 088 West Bengal 033-24505550/ 54 033-24392704 bltea@cal3.vsnl.net.in
Weikfield Products Co. (India) Pvt. Ltd., Weikfield Estate, Nagar Road, Pune- 411 014 Maharashtra Tel: 020 26633111/2 020 2663 3380 Email: weikfield@weikfield.com	The United Nilgiri Tea Estates Co. Ltd., (UNTE) Chamraj Estate P.O.643 204, Nilgiris, Tamil Nadu. Tel: 0423 225 8737 0423 225 8837 Email: chamraj@vsnl.com	Tea Group Exports 20, Coal Berth, Hoboken Road, Kolkata 700042, West Bengal Tel: 033 24391966 Email: ambootia@vsnl.com	Parry Agro Industries Limited Iyerpadi Estate, Iyerpadi P.O. 642 108, Via Pollachi Coimbatore Dist. Tamil Nadu Tel: 91-4253-222489 / 564 91-4253-222250 Email: muralipadikkal@pai.murugappa.com
Tata Tea Limited, Tata Tetley Division, Mr. K. M. Angelos 73/ 74, K.P.K.Menon Road, Willingdon Island, Kochi-682003 Kerala Tel: 0484-2667427, 2668758, Mob: 09895712001 484-2666808 Email: tetley@md2.vsnl.net.in	Godfrey Phillips India Limited 3 Cooper Street, 1st Floor,Kalighat P.O. Kolkata-700 026 West Bengal Tel:033-2486 0178/179 033-2486 0177 Fax: 033-2486 0177 Email: akstea@vsnl.net	Parry Nutraceuticals Limited 43, moore St. Parry House, 5th Floor Chennai – 600 001 044-25036816 SebastianT@pai.murugappa.com	Doon Heights Agro 260, Phase I, Vasant Vihar, Dehradun – 248 006 Uttranchal Tel: 91 135 2763620 Fax:91 011 29216352 Email: ranjitlall@doonheightsagro.com, ranjit_lall@yahoo.com

Name and Address of Exporter	Name and Address of Exporter	Name and Address of Exporter	Name and Address of Exporter
Harrisons Malayalam Limited Touramulla Organic Division, Chundale Estate, Touramulla P.O., Wayanad 673 592 Kerala Tel: 04936 202688 chundale@eth.net	Golden Mist Plantations/Birhat Consultants (I) Pvt. Ltd. Golden Mist Plantations & Resorts Pvt. Ltd., Galibeedu Village, Mercara, Kodagu, 571 201 Karnataka Tel: 08272 265 629 ludwigorganic@hotmail.com	Mr. Paras Desai Gujarat Tea Processors & Packers Ltd. Waghbakri House, Opp. Parimal Garden Ambawadi, Ahmedabad 380 006 Gujarat	Bush Tea Co., Pvt. Ltd 18, A, Park Street, Stephen Court (6th floor), Kolkata West Bengal Tel: 033 22296705/06 033 22496707
Mr. Cherian Xavier Accelerated Freeze Drying Co. Ltd Amalgam House, Bristow Road Willingdon Island Cochin – 682003, Kerala	Mr. B.D. Agarwal Vikas Wsp Ltd. (Adm. Office), B-86-87 Udyog Vihar, RIICO Industial Area, Sri Ganganagar-334002 (Rajasthan) Tel: 91 154 2494512/52 Fax: 91 154 2494361	Mr. Ajay Katayal Sunstar - Organic Food Division, 40 km. Stone, G.T. Karnal Road, Bahalgarh-Sonepat, Haryana Tel: 0130-2381155, 2381925, 2381926 Fax: 0130-2381055	Mr. Mayank S. Patel Urvesh Psyllium Industries Ltd., State High Way, Nr. Khali Char Rasta, Khali, Sidhpur-384151
Elk Hill Organic Project Post Box No. 12, Sidapur 571 253, Coorg Karnataka Tel: 08274 267756/50 08274 258368 Email: fairland@sancharnet.in	Mr. Ajay Kapoor S & D Aroma (India) Pvt. Ltd. C 5/6 LGF Grand Vasat, Vasant Kunj, New Delhi 110070 Tel: 09818010101	Achal Industries 190, Industrial Area, Baikampady, Mangalore 575 011 Karnataka Tel:0824 240 8187 0824 240 8487 Email: achalind@vasnet.co.in	Mr. Jagajit Singh Little Bee Impex G.T. Road, Doraha, Punjab 141421 Punjab Tel: 01628 258240/258640 01628 259570/258140 Email: export1@kashmirhoney.com

Name and Address of Exporter	Name and Address of Exporter	Name and Address of Exporter	Name and Address of Exporter
Col. Deepak Badhwar Clestia Impex Pvt. Ltd Flat 557, Tower III, Kailash Apartments, East of Kailash, New Delhi 110065 Tel: 011 26483059, 0135 2742433/3958955 011 262225293 Email: dbadhwar@rediffmail.com	Mr. Ravinder Kumar Bhandora Organics 29, Sant Nagar, II nd floor East of Kailash, New Delhi 65 Tel: 011 26433934 011 26433934 Email: info@bhandora.com	Mr. Jose Dominic Natural Harvest (India) Pvt. Ltd. XXIV/1350, Casino Hotel, Willindon Island P.O. Cochin 682 003 Kerala Tel: 0484 2668221 0484 2668001 Email: naturalharvest@gmail.com	Mr. Vimal Anand Apis (India) Natural Products 18/32, East Patel Nagar, New Delhi – 110 008 Tel: 011 25737038 Email: vimal@apisindia.com, mail@apisindia.com
Rani Tea Estate/ MKB Asia (P) Ltd P.O. Rani, Guwahati 781 017, Kamrup Dist Assam Tel: 0361 284 2059, 284 0074 0361 241 7085	Muscatel Valley Muscatel Valley Division of Goomtee Tea Estate, P.O. Mahanadi 734 223, Darjeeling West Bengal Tel: 0354 233 8022 0354 233 8011 E-mail: slg_goomtee@sancharnet.in	Satnam Overseas Ltd; 50-51, K.M. Stone, G. T. Road, Murthal – 131027, Sonepat Haryana Tel: 0130 2482043, 2482456 E-mail: satnamoverseas@vsnl.com	Kurinji Organic Foods (India) Pvt. Ltd. Periyakulam Road, Genguvarapatti, Theni Dist, 625 203 Tamil Nadu Tel: 04543 262469/263 575 04543 265 496 Email: jakes@md3.vsnl.net.in

Name and Address of Exporter	Name and Address of Exporter	Name and Address of Exporter	Name and Address of Exporter
Thiashola Estate Thiashola Post, The Nilgiris, 643 230 Tamil Nadu Tel: 0423 250 9897/2509244 Email: Thiasholaestate.Hll@unilever.com	Tata Tea Limited Milne Road, Willingon Island, Kochi 682 003 Kerala Tel: 0484 2668356 0484 2668076 Email: raveendran.ma@tatatea.co.in	Sutlej Power Private Limited 610 – 611, Prakash Deep Building, Tolstoy Marg, Connought Place, New Delhi 110 001 Tel: 011 51520078/80, 3322560/63 011 23322542 Email: sutlej_organics@yahoo.co.in	Tata Coffee Limited Pollibetta, South Kodagu, 571 215 Karnataka Tel: 08274 251411/12/13/82/83 08274 251425 Email: hh@tatacoffee.com
Mr. Karan Singh Jamwal Kittu Exports Kotla Mubarkpur 110003, No-1 1st Floor Sewa Nagar Mkt. New Delhi	Mr. K.B. Singh Amar Singh and Sons, 51 - Industrial Estate, Phase - 1, Gangyal, Jammu Tawi - 180010 Jammu & Kashmir Tel: 0191-2480606, 2480350, 2480975 Fax: 0191-2480568, 2480473	Sonarie Tea Estae & MRB Enterprises Jonak Division, P.O. Sonarie, Shivasagar – 785 690 Assam Tel: 03772 256578 0361 220 2746	Welbeck Tea Estate Lovedale SO, Udhagamandalam, 643 003 Tamil Nadu Tel: 0422 2212659 0422 2218628 Email: welbecktea@sancharnet.in
Mr Rohit Doshi Sankalp Bio Cotton project, 202, kuber house, 162, Anchan bag, Indore: 452001	Mr Mahesh H. Thakker Mahesh Agri Exim Pvt. Ltd. 312, Sharda Chamber No. 1, 3rd Floor, 31, Keshavi Naik Road, Bhat Bazar, Mumbai – 400 009	Mr Rohit Doshi Swayam Bio Cotton Project 202, Kuber House, 162, Kancahan Bag, Indore 452 001	Mr. Dilip Singh Rathore Rathore Organic Farming Village Post - Bhilgaon Tehsil – Kasrawad, Dist. Khargone, M. P.

Name and Address of Exporter	Name and Address of Exporter	Name and Address of Exporter	Name and Address of Exporter
Mr. Sandhir Agarwal Kamala Tea Company Ltd 240/B, 3rd Floor, A. J. C. Bose Road, Kolkata – 700 020	Mr. Jyotindra Jyotindra International 4 Km, Palanpur Ahmedabad Highway Palanpur 385 001	Mr. Mohan Lal Agarwal Kanchanjunga Herbal Medi Aids Pvt. Ltd. 669, Marshal House, 33/1, Netaji Subhas Road Kolkata-700 001, West Bengal	Mr. Srinivasa M. Rao ITC Ltd – International Business Division 31, Sarojini Devi Road Secunderabad – 500003 Andhra Pradesh
Mr. Atul Kumar Artousi Enterprises B-26, FreedomFighters Enclave, Neb Sarai, New Delhi-110 068 Tel 011-26566611 Fax: 011-26530789 E-mail :artousi@rediffmail.com	Mr.Siddharth .S.P.Zantye Amruta Cashew Industries Nathpai Road, Vengurla, Sindhudrg Dist Maharashtra	Mr. Sanjay Udashi Shree Sanjay Trading Company A/7, Majithia Appartments, S.V. Road, IRLA, Vile Parle (West) Mumbai	Ajanta Industries Plot D, Sur. No. 166/1, Apewal Ponda, Goa-403401
Mr. Laxman Prabhu Damodar Cashew Company Kotachery, Kanhangad – 671 315 Kerala	Mr. Amir Ahasan Assam Company Ltd. Post Salonah, Nagaon District, Assam	Mr. Dharmesh Patel Geo Fresh Organic Gulab Park, S.T.Road, Sidhpur-384151, Gujarat	Mr. Saurab Garg Aryan International D – 184, Freedom Fighters Enclave Neb Sarai, New Delhi – 110068
Mr. Jayaprakash Narakkat Indway International Asia Fruits, 14 Skandha, Pestom Sagar Road 2, Mumbai	IITC Organic India Pvt. Ltd. Village Kamta, Post Chinhat, Faizabad Road	Mr. Danesh Madon Eaternal Health & Organic Foods Pvt. Ltd. B/2, Rustom Baug, S. S. Marg, Byculla, Mumbai-400 027	Mr. Udaya Kumar Hope Spice Project No 51 Murugan lodge, Ettines Road, Ooty 643002, Tamil Nadu
Mr. Umesh N. P. Zantye Hira Cashew Industres Vithalapur,Sanquelim Goa State	Mr. Kishore Bheda H.Bheda &Co. 202 kapadia Apartments 39,S.V. Road, Vile Parle (W) Mumbai – 400056	Mr. Jai Chaitanya Dasa Eco Agri Research Foundation 142, 4th Main Road, 4th Cross, Bannimantap ‘C’ Layout Mysore, Karnataka State	Mr. Samir Changoiwala Gopaldhara International 17, Ganesh Chandra Avenue, Kolkata-700013, West Bengal

Name and Address of Exporter	Name and Address of Exporter	Name and Address of Exporter	Name and Address of Exporter
Mr. Swaraj Kumar Banerjee Makaibari Tea Estates Kurseong, Darjeeling District West Bengal-734 203	Mr. G. Krishnan Nair Krishnan Food Processors P. B. No. 344, Iyshwaria Beach Road, Kollam-691 001, Kerala	Mr. Sudheer Mudar Mudar India Exports 6 – 426 – C2, Kovoor Nagar, Anantapur. Andhra Pradesh	Mr. Anil Kr. Mittal KRBL Limited A1, Pamposh Enclave New Delhi-110 048
Sardar Patel Farm Dr. Dinesh Patel Naroda Kathwada Road, Kathwada, T. Daskoi, Ahmedabad.	Phaladaayi Foundation Shobha N. Shastry No. 266, 7th Cross, 9th Main, Ideal Homes Township, Bangalore - 560 098	Ind Frag Mr. K. K. Sreeram S. No.102/3,12th K. M. Hosur- Kelamangalam Road, Kundumaranapalli, Tamilnadu 635113	Organic Farms, Near Jain Temple, Sawkarpura, At Post. Anjangaon SurjiDist. Amravati – 444 705, Maharashtra State TEL: +91 7224 242707
Fr. Augustine Kariapuram Peermade Development society P. O. Box 11, Peermade Idukki 685 531 Kerala	Mr Uday Singh Namdhari Fresh Bidadi – 562 109, Bangalore, Karnataka	Ms. Shobha N. Shastry Phalada Agro Research Foundations Pvt Ltd No. 266, 7th Cross, 9th Main, Ideal Homes Township, Bangalore - 560 098	Suresh N. P. Zantye N.G.P.Zantye & Co Bicholim . Goa - 403 504
Fr. Kuriakose Kunnath Wayanad Social Service Society Post Box No. 16, Mananthavady, Wayanad-670 645, Kerala	Mr. N. Viswanathan Vishwas Organic Technologies "Lahari", No. 83, 14th Main, 4th Sector, HSR Layout Bangalore-560 034	Mr. Mukesh Malhotra Weikfield Products Company Pvt. Ltd. Weikfield Estate, Nagar Road, Pune-411 014, Maharashtra Tel: 020 – 26633111/12 Fax: 020 – 26633380 E-mail: weikefield@weikfield. com	Mr. Nirmal Mehta Veetee Fine Foods Ltd. Veetee House, 56-57 KM, G.T. Karnal Road, Larsauli District Sonepat - 131 001

Name and Address of Exporter	Name and Address of Exporter	Name and Address of Exporter	Name and Address of Exporter
Mr. Sameer Mehra, Suminter India Organics 308, Oberoy chambers,-1, New Ring Andheri Mumbai	Mr. Murali B. N Sri Mata Bio Source No. 9/5, 16th Cross, Indian Bank Colony, Raja Rajeshwari Nagar, Bangalore – 560 098 Tel: 0091-80-26606259, Fax: 26697992	Mr. Binod Mohan Tea Promoters (India) Pvt. Ltd 17, Chowringhee Mansions 30, Jawaharlal Nehru Road Kolkotta – 700016	Mr. Rajashekar Reddy Seelam Managing Director Sresta Natural Bio Products Pvt. Ltd Sresta House Plot # 7, LIC Colony, Sikh Village, Secunderabad – 500 009 Tel: +91 40 27893028, Fax: + 91 40 27893029 Mobile: +91 0 9392493631 E-mail: rajseelam@sresta.com www.sresta.com
Mr. Sameer Azad Kashmir Kessar Mart P.O. Box No. 216, Munwarabad, Srinagar Jammu & kashmir Tel 0194-2470846, 2470932 Fax 0194-2473169 kkmartsqr@hotmail.com	Mr. Raja Singh Kapoor Kashmir Apiaries Village: Mallipur, G.T.Road Doraha-1410421, Dist: Ludhiana, Punjab	Mr. Vikas Bhat Khandige Herbs & Plantations Pvt. Ltd No. 46/1, Jaraganahalli, Kanakapura Main Road Bangalore - 560 045	Mr. P. K. Panigrahy Kasam Kandhamal Apex Spices Contractorpada, Gurumurthy road Phulbani, Orissa – 762 001 Tel: 06842- 253022, 253531, 253380

Name and Address of Exporter	Name and Address of Exporter	Name and Address of Exporter	Name and Address of Exporter
Mr. Tomy Mathew Elements Homestead Products Private Limited 4/1418-B Customs Road, Calicut, Kerala 673032 Ph: 95 495-2765783 (O), 2368153 (R) Email: elements@dataone.in, tomy@elementsindia.net	Mr. Suresh N. P. Zantye Zantye Cashew Industries Dhuriwada, Malvan, Sindhudurg District, Maharashtra	Mr. Bijumon Kurian Manarcadu Social service Society Mini Industrial Estate Manarcade P.O Kottayam, Kerala Ph:0481-6532878, 9447066757/2 masss@sify.com, plantrich@plantrich.com	Mr. Joji Farmer Industries (Renamed as Golden Vintage Farmers Industry) Khanna Nagar P.O, Muringoor, Chalakudy Trissur, Kerala – 680309 Ph:0480-2709618,2709718 Email:sales@farmerindia. commm@farmerindia.com, ceo@farmerindia.com
Mr. Sayeed A. Patel Rajena Agro Products Pvt. Ltd Sedrana, Sidhpur, Patan-384 151 North Gujarat	Mr. Mohan Chirimar Raghunath Exporters 5F, Park Plaza, 71, Park Street, Calcutta – 700 016 Tel: 033 – 22290092/22294495	Mr. A. Somasundaram SNV Horticultural Farms Savadippatti Village, Alagapuri – 625 523, Theni District, Tamil Nadu	Mr. Thomas Jacob Poabs Exports Poabs Organic Products Pvt. Ltd. Seethargundu PO Nelliyampathy Palghat, Kerala India – 678511 Tel: 0492 – 346554, 346333, 346223 Telfax: - (0492) 346555 E –mail: mail@poabsorganic.com

Source: http://agritech.tnau.ac.in/org_farm/orgfarm_exporters.html

Annexure-VI

Organic Importers

Name and Address of Importers	Name and Address of Importers	Name and Address of Importers	Name and Address of Importers
Jurg Tumbrunnmen Surgo Ag, Basel, CH Tel : + 4161 317 3253	Mr. Anil Puri A.M Stubacher Berg-10 D-91468 Gutenstetten Germany Fax No.: 0421-2475394	Mr. Anil Puri A.M Stubacher Berg-10 D-91468 Gutenstetten Germany Fax No.: 0421-2475394	Ian Taylor Fuerst Day Lawson Ltd St Clare House 30-33 Minories London EC3N 1 LN Tel: +44171 4880777 Fax:+44171 702 1500
Germany Agasaat Gmbh Gewerbegebiet Sued, Pascalstr.11. D-47506,Neukirchen- Vluyn,Germany Tele: +49 (0) 28459146-0 Fax: +49(0) 28457573 Internet: www.Agasaat.De	Bioprim Address:530 Av De Milan,Zi Du Grand Saint Charles 66000 Perpignan France Tel: +33(0) 4 6854 7979 Fax: +33(0) 4 6758 5970 E-mail: Contact@Bioprim.Com	Dynamis France 54 Avenue De La Vilette,94637 Rungis Cedex France Tel: +33 (0) 1 4560 4344 Fax: +33(0) 1 4687 4405 E-mail: Dynamis@Wabdii,Fr	Denree Versirgungs Gmbh Hofer Str 11 D-95183 Topen,Germany Tel: +49(0) 9295 180 Fax: +49(0)9295 1850 E-mail: Webernk@T-Online.De
El-Ouente Gmbh Hildesheimestr,59 D-31177, Harsum Germany Tele: +49(0) 5127-98860-0 Fax: +49(0) 5127-9886028 E-mail: Info@El-Puente.De Internet: www.El-Puente.De	Naturkost Schramm Ludwig-Winter- Strasse 6 D-77767 Appenweier Germany Tel: + 49 (0) 7805 98880 Fax: + 49(0) 7805 966880 E-mail: Team@Naturkost- Schramm.De Internet: www.Naturkost-Schramm. De	Voelkel Gmbh Fahrstr.1 29478,Hohbeck/Ot Pevestorf Germany Tel: +49 (0) 5846 9 50-0 Fax : +49(0) 5846 9 50 50 E-mail: Marketing @Voelkeljuice. De Internet:www.Voelkeljuice.De	Ctm Altromercato Via Macello,18 39100 Bolzano Italy Tel: +39 (0) 471 975 333 Fax: +39(0)471 977 599 E-mail: Info@Altromercato.It Internet:www.Altromercato.It

Name and Address of Importers	Name and Address of Importers	Name and Address of Importers	Name and Address of Importers
Loders Croklaan P.O. Box 4, 1520 Aa Wormerveer The Netherlands Telephone: +31 (0) 75 629 2911 Fax: +31 (0) 75 628 9455 E-mail: Fats.Lc@Croklaan.Com Internet: www.Croklaan.Com	The Nertherlands Platinastraat 50 8211 Ar Lelystad, The Netherlands Tele: +31(0) 320 282 928 Fax: +31(0) 320 282 028 E-mail: Info@ Detraay.Com	Neuteboom B.V. Aadijk 41, 7202pp Almelo The Netherlands Telephone : +31 (0) 546 864 062 Fax: +31 (0) 546 866 369 E-mail: Info@Neuteboom.Nl Internet www.Neuteboom.Nl	Rhumveld Winter & Konijn Bv P.O. Box 29216, 30001 Ge Rotterdam The Netherlands Telephone: +31 (0) 10 233 0900 Fax: +31 (0) 10 233 0574 E-mail: Rwk@Rhumveld.Com Internet: www.Rhumveld.Com
Ernst Weber Naturkost Postfach 750954 D-81339 Munchen Germany Tel: +49(0) 89 746 3420 Fax: +49(0) 89 746 342222 E-mail: Webernk@T-Online.De	Exodom P O Box 7025 69348,Lyon Cedex 07 France Tel: +33(0) 4 3728 7350 Fax: +33(0) 4 3728 7354 E-mail: Exo-Dom@Wanadoo.Fr Internet: www.Exodom.Com	Image Tutti Verde Marche Saint Charles P O Box 5129,66031 Perpignan France Tel: + 33(0) 4 6868 4040 Fax: +33(0) 4 4868 4048 E-mail: Imago1@Wanadoo.Fr	Gepa Fair Handelshaus Gewerbepark Wagner Bruch 4 42279 Wuppertal Germany Tel: +49 (0) 202 266 830 Fax: +49(0) 202 266 8310 E-mail: Marketing@Gepa.Org Internet: www.Gepa3.De
WalterDanzer Turmstr.6 8952 Schlieren Tel: 01 731 1200 Fax: 01 731 1275	Dr Metz KG Siemenstr. 7 65779 Kelkheim Tel: 06195 3071/3072 Fax: 06195 8729	Galke Hartmut Am Bahnhof 1-5 D- 37534 Gittelde Tel: 05327 868 10 Fax: 05327 5420	Rapunzel Naturkost AG Haldergasse 9 D-87764 LEGAU Tel: 083 30910 –150 Fax: 083 30910 175

Name and Address of Importers	Name and Address of Importers	Name and Address of Importers	Name and Address of Importers
Heuschrecke Naturkost Gmbh Redcastr, 50a 53842,Troisdorf-Spich Germany Tel: +49(0) 2241-39726-0 Fax: +49(0)2241-39726-99 E-mail: Bio@Heuschrecke.	Spack Bv Telephone: +31 (0) 181 48 6486 Fax: +31 (0) 181 48 6857 E-mail: Spack @Xs4all.Nl	Simon Levelt B.V. A.Hofmanweg 3, 2031 Bh Haarlem Telephone: +31 (0) 23 512 2522 Fax: +31 (0) 23 512 2505 E-mail Info@Simonlevelt.Com Internet: www.Simonlevelt.Nl	Uni-Vert Route De Bellegarde 30129, Manduel, France Tel: +33(0) 4 6620 7525 Fax: +33(0) 4 6620 7526 E.mail: Uni-Vert@Uni-Vert.Com Internet: www.Uni-Vert.Com
Danner, Mr Weinzier Labertalstr. 4 D-93161 Alling/Regensburg Tel: 49 9404 95 55 Fax: 49 9404 2096	Integgerij G.a. van der Kroon b.v. Van Heemskerckstraat 31 Postbus 17 4670 AA dinteloord (Holland) Tel: 01671 52 22 50 Fax: 01671 52 30 05	Tradin Bv Latexweg 12, 1047 Bj Amsterdam Telephone: +31 (0) 20 407 4499 Fax: +31 (0) 20 497 2100 E-mail: Info@Tradinorganic.Com Internet: www.Tradinorganic.Com	Programme Manager CBI, Organic Food Ingredients 3011 AA Rotterdam Beursplain 37 The Netherlands
Alterbio France Sarl 5 Rue Levasseur,Zi Saint Charles, 66000 Perpignan, France Tel: +33(0) 4 6868 3838 Fax: +33(0)4 6868 3829 E-mail: Info@Alterbio.Com Internet: www.Alterbio.Com	Arcadie 5 Rue Levasseur, Zi Saint Charles 66000 Perpignan, France Tel: + 33(0)4 6656 9933 Fax: +33(0) 4 6630 6261 E-mail: Info@Arcadie-Sa.Fr Internet: www.Arcadie-Sa.Fr		

Source: http://agritech.tnau.ac.in/org_farm/orgfarm_importers.html